THE NUCLEAR FUEL CYCLE

A Survey of the Public Health, Environmental, And National Security Effects of Nuclear Power

The MIT Press Environmental Studies Series
Gordon J. F. MacDonald, consulting editor

Man, Materials, and Environment, National Academy of Sciences and National Academy of Engineering, 1973

Citizen Groups and the Nuclear Power Controversy: Uses of Scientific and Technological Information, Steven Ebbin and Raphael Kasper, 1974

The Nuclear Fuel Cycle: A Survey of the Public Health, Environmental, and National Security Effects of Nuclear Power, Union of Concerned Scientists, 1975

THE NUCLEAR FUEL CYCLE

A Survey of the Public Health, Environmental, and National Security Effects of Nuclear Power

Revised Edition

Prepared by
The Union of Concerned Scientists

The MIT Press
Cambridge, Massachusetts, and London, England

This book was typed by Sheila M. Toomey, printed on Finch Publisher's Offset by Halliday Lithograph Corporation, and bound in MBV-4259 by the Colonial Press Inc., in the United States of America.

Library of Congress Cataloging in Publication Data

Union of Concerned Scientists.
The nuclear fuel cycle.

(MIT Press environmental studies series)
Includes index.
1. Atomic power--Addresses, essays, lectures [hazards] . 2. Atomic energy industries--Addresses, essays, lectures [hazards] . I. Title.
TK9155.U54 1975 614.8'39 75-25759
ISBN 0-262-21005-3 (hardcover)
ISBN 0-262-71003-X (paperback)

CONTENTS

THE AUTHORS

Daniel F. Ford
Executive Director,
Union of Concerned Scientists

Philip M. Hatfield, MD
Radiologist

Thomas C. Hollocher
Associate Professor of Biochemistry,
Brandeis University

Henry W. Kendall
Professor of Physics
Massachusetts Institute of Technology

James J. MacKenzie
Joint Scientific Staff,
Massachusetts and National Audubon Societies

Marc Ross
Professor of Physics,
University of Michigan

Lawrence Scheinman
Professor of Government, and
Associate Director of Peace Studies Program,
Cornell University

Arell S. Schurgin
Environmental Studies Program,
Brandeis University

Nuclear power plants, Indian Point, Westchester County, New York. At the time this photograph was taken, one plant was in operation and two were still under construction. UCS photo, December 1973.

PREFACE

A worldwide controversy has developed over the safety, environmental, and national security impact of the expanding program of utilizing nuclear reactors to generate electricity. The potential hazards of nuclear power plants are the subject of debate because of the very large quantities of radioactive materials inevitably associated with their operation, and because nuclear power plants produce materials from which nuclear explosives can be made. There are a number of opportunities during the nuclear fuel cycle (i.e., during the mining, processing, transportation, utilization, and storage of nuclear materials) where significant radioactive exposures or other safety hazards to the public may arise. Some of these problem areas, such as those associated with the long-term storage of high-level radioactive wastes, require solutions which will endure for hundreds of thousands of years. In this sense, and for this reason, the problems of handling nuclear materials are unique and deserve the most careful attention.

An official U.S. Atomic Energy Commission (AEC) assessment of the hazards associated with the handling of nuclear materials, entitled "Environmental Survey of the Nuclear Fuel Cycle," was published in November 1972. The survey dealt with uranium mining, milling, and enrichment; fuel fabrication and reprocessing; and the transportation and storage of radioactive wastes. It did not touch upon the problems of accidental releases of radioactivity from reactors, or the safeguarding of nuclear materials to prevent their unauthorized use in the fabrication of nuclear weapons.

Following the release of the AEC's assessment of the risks associated with the nuclear fuel cycle, the Union of Concerned Scientists (UCS) initiated an independent, parallel study intended to illuminate defects and important omissions in the AEC survey. Studies were required on essentially all of the topics mentioned above. This book presents the conclusions of the major technical review that was carried out. It is the most recent in a series of UCS studies on such topics as fossil power plant air pollution, liquefied natural gas transport and storage hazards, oil spill analyses, and a series of reports on the nature and chances of catastrophic nuclear reactor accidents.

The book contains reviews of the exposure of uranium miners to radioactive gas; the milling of uranium and the subsequent disposal of radioactive tailings; the possibility, mechanisms, and consequences of catastrophic accidents; problems of diversion, during routine transport, of nuclear material for terrorist purposes; the hazards of transporting radioactive materials; some of the operating history of the first U.S. fuel reprocessing plant; and, finally, the storage and disposal of high-level radioactive wastes.

One issue relating to the nuclear fuel cycle is omitted here: public health problems related to the routine radioactive releases from nuclear power plants. This matter has been the subject of a thorough, independent investigation[1] which needs no duplication.

A preliminary version of this book was published in 1974, under the same title, for the UCS by Friends of the Earth.

The individual chapters of this book were

reviewed by their respective authors for accuracy. Responsibility for the content of each chapter rests with them.

Reference is made throughout this volume to the U.S. Atomic Energy Commission (AEC). It should be noted that this agency was abolished by the Energy Reorganization Act of 1974, which went into effect on January 19, 1975. The Regulatory Staff of the AEC, under five Commissioners appointed by the President, became the Nuclear Regulatory Commission; the bulk of the AEC was incorporated into the Energy Research and Development Administration under a single Administrator, also appointed by the President.

1 "The Effects on Populations of Exposure to Low Levels of Ionizing Radiation." Report of the Advisory Committee on the Biological Effects of Ionizing Radiations, National Academy of Sciences - National Research Council, November 1972.

ACKNOWLEDGMENTS

The Union of Concerned Scientists is grateful to the many people who have helped with and supported the preparation of this book, among them the authors, each of whom has contributed his time in the public interest.

We acknowledge with grateful thanks support from the Bydale Foundation, the Stern Family Fund, the Ottinger Foundation, and Mrs. Lucy B. Lemann. We appreciate the help of Mr. Anthony Z. Roisman, of Berlin, Roisman and Kessler, Inc., for his coordination of this project and for much other aid. We are grateful for substantial help in the preparation of the manuscript from Ms. Regina E. Ford and Ms. Joan Murphy.

UCS Steering Committee

Daniel F. Ford
Henry W. Kendall
James J. MacKenzie

DEFINITIONS OF TERMS

Abbreviations of the Elements. Many elements are referred to in this report. Some of the more frequently cited ones, along with their chemical symbols, are: strontium (Sr), cesium (Cs), iodine (I), krypton (Kr), potassium (K). Other elements are listed below under Actinides.

Actinides. The Actinides are those elements in the periodic table beginning with and following actinium: actinium (Ac), thorium (Th), protactinium (Pa), uranium (U), neptunium (Np), plutonium (Pu), americium (Am), curium (Cm), plus several other very rare elements.

Alpha Wastes. Radioactive materials which emit alpha particles, i.e., helium nuclei, as they disintegrate. Alpha wastes are dangerous if inhaled or ingested. (Alpha is represented by the symbol α.)

Biological Half Life. The time required for a biological system, such as a man or an animal, to eliminate, by natural processes, half the amount of a substance (such as a radioactive material) that has entered it.

Curie. The unit describing the radioactivity of a sample of material. A curie (abbreviated Ci) is equal to 37 billion (37×10^9) disintegrations per second.

Fission. The splitting of a nucleus into two or more parts. These resulting parts (fission products) are usually very radioactive.

Gw(e) or Gigawatt (Electric). One billion (10^9) watts of electric generating capacity.

Half Life. The time in which half the atoms of a particular radioactive substance disintegrate to another nuclear form.

Man-Rem. The product of the average individual dose (in rems) in a population times the number of individuals in the population.

Metric Ton. 2,205 pounds.

MeV or Million Electron Volts. A unit of energy used to describe radioactive emissions. It takes about 23 billion billion MeV (23×10^{18}) to make one kilowatt-hour.

Mw(e) or Megawatt (Electric). One million (10^6) watts of electric generating capacity.

Order of Magnitude. A factor of ten. Thus, for example, three orders of magnitude means a factor of 1,000.

Rad or Radiation Absorbed Dose. The basic unit of absorbed dose of ionizing radiation. A dose of one rad means the absorption of 100 ergs of radiation energy (about 2.7×10^{-12} kilowatt-hours of energy) per gram of absorbing material.

Rem or Roentgen Equivalent Man. A unit of dose of radiation. Common doses received from background radiation are on the order of 0.150 rem/yr. A millirem (mrem) is one thousandth of a rem.

Transuranic, Transuranium. Pertaining to those elements following Uranium in the periodic table - usually the first four, Np, Pu, Am, Cm.

THE NUCLEAR FUEL CYCLE

A Survey of the Public Health, Environmental, And National Security Effects of Nuclear Power

Chapter 1

AN OVERVIEW OF NUCLEAR POWER

Daniel F. Ford, Henry W. Kendall, James J. MacKenzie

The United States is presently deeply involved in an issue of fundamental significance - the "energy crisis." The issue is a tangle of problems relating both to immediate and to long-term availability of energy and to the impact of energy production and utilization on human health, the environment, and national security. Attempts at resolution of the array of difficulties raises a new set of problems: what new sources of energy are available to mitigate expected shortages? Which are feasible? And of those, which are acceptable? Should the country begin increasingly to conserve energy? How can this be done? And how effective might it be? Such questions are now of critical importance to the energy issue.

At present, the nation has no orderly means of obtaining the assessments and solutions - frequently highly technical - required to deal with the energy crisis. One especially thorny difficulty is presented by the electric supply problems of the country and the role that nuclear power should play in alleviating them.

The electric power needs of the United States have been doubling approximately each ten years for over seven decades. The generating capacity to meet these needs has, until recently, been made readily available. Within the past few years, however, an array of difficulties has accumulated that increasingly is restricting both the development of the generating capacity and, in some cases, its utilization, in amounts adequate to meet the demands. Among these difficulties are acute shortages of fuels and increasing environmental constraints. The shortages, especially of natural gas and oil, have been induced by a combination of factors including federal pricing policies over natural gas; a peaking in domestic crude oil production; and troubles in international supply. In addition, uncertainties in the supply of crude oil and in federal energy policy have made domestic refinery construction unattractive to the major oil companies. Coal, representing our largest fossil fuel reserves, faces well known and formidable problems in both mining and burning.

Nuclear generated electric power, recognized as a technical possibility from the time of the Manhattan Project in World War II, has been heralded as a power source that could relieve the country of its electrical supply troubles. Indeed, it was believed that power could be generated from the nucleus in safe, environmentally acceptable ways at such low cost that it would be "too cheap to meter." A program to implement this goal has been underway under the auspices of the U.S. Atomic Energy Commission since the middle 1950s with the initiation of construction of the first civilian nuclear power installation, the Shippingport reactor in Pennsylvania. When this nuclear power plant went "on-line" in 1958 it was the first in a series that now comprises 56 operable power reactors (supplying 7% of the nation's electricity), 60 more under construction, and 105 ordered. Operable plants have an installed capacity in excess of 30,000 Mw(e). Those plants ordered or under construction represent a potential capacity of 173,918 Mw(e). According to a recent projection of the Department of Interior,[1] by 1985 nuclear capacity will reach 215,000 Mw(e), and by the end of the century, 960,000 Mw(e). The AEC has made even more optimistic projections. By 1985 the AEC expects a nuclear capacity of 280,000 Mw(e), and by the year 2000, 1,200,000 Mw(e).[2]

If the more modest projection of the Department of Interior is realized, nuclear power plants will supply about 26% of total U.S. energy needs by the turn of the century, representing 61% of our supply of electricity. Even now, New England depends on nuclear reactors for 20% of its electricity, and by 1982 this dependence is expected to rise to about 40%.

Nuclear power has characteristics strikingly different from any other technology that has been developed and implemented by industrialized nations. These characteristics are related to the radioactivity, biologically highly toxic, which is inevitably associated with the production of electricity by nuclear means and with the fissionable materials used as fuels. The radioactivity accumulates in immense quantities, and, owing to its biological effects, has the potential for causing damage on a large scale. The fissionable materials may, under some circumstances, be employed to produce nuclear explosives. The major safety issues related to nuclear power production can conveniently be divided into three categories. The first has to do with the possibility of a reactor accident involving the release of an extremely large quantity of radioactivity. Such an accident could have catastrophic consequences. The second area of concern is the problem of the transportation, processing, and disposal of the radioactive "ashes" produced in nuclear reactors. These materials must be separated from the environment for nearly geological periods of time. The third area relates to the problems of safeguarding certain nuclear materials so that they cannot be diverted to terrorist uses. In addition, there are public health and safety problems related to the mining and milling of nuclear fuels.

In the first two major areas there are two sorts of problems. First, there are important technical problems of understanding how to control and contain the radioactivity under all conceivable circumstances, both in the course of reactor operation and in the transporting, processing, and disposal of material after it is removed from the reactors. Additionally, however, there are societal problems related to the difficulties of providing the stringent management required in all facets of a nuclear program. These are especially severe in light of the potential for great harm should things go wrong. These management problems are common also to the safeguards area.

If the United States continues to move toward a heavy reliance on nuclear power, it should recognize that it is adopting a technology with the potential to cause spectacular harm. It must recognize that new, and quite extraordinary controls throughout the program are required to prevent accidents of unparalleled scale from occurring in the operation of large numbers of power reactors. A thoughtful proponent of nuclear energy, Dr. Alvin Weinberg, formerly Director of the AEC's Oak Ridge National Laboratory, has written:

"We nuclear people have made a Faustian compact with society: we offer an almost unique possibility for a technologically abundant world for the on-coming billions, through our miraculous, inexhaustible energy source; but this energy source at the same time is tainted with potential side effects that, if uncontrolled, could spell disaster."[3]

The potential consequences of a reactor accident, of a successful detonation of a terrorist's nuclear weapon, or of the unintended release of a large amount of radioactive material from archival storage to the environment establish the level of care, diligence, and understanding required in the management and operation of a nuclear program. "The price we demand of society," Weinberg continues, "for this magical energy source is both a vigilance and a longevity of our social institutions that we are quite unaccustomed to....Is mankind prepared to extend the eternal vigilance needed?"[4]

That this compelling need for diligence is not always fully perceived is illustrated in the U.S. program. Nuclear power plants are surrounded by controversy concerning an array of safety related problems that are at once unique and of critical importance. The problems have now become the source of deep concern to many scientists in the United States and elsewhere, because some of the most important of them are unresolved. As a consequence of this concern there is now increasing public questioning about the role the nucleus should play in contributing to the solution of the country's energy problems. Weaknesses have been identified in the capacity of reactor emergency systems and, thus, in the assurances of security against catastrophic accidents. Moreover, there has been loss of management control over plant quality assurance and over design, maintenance, and operating procedures. As a consequence, the program has, according to an AEC task force report, been "besieged" by safety problems. The AEC's materials—safeguards program has been labeled "inadequate" by all outside review groups who have inspected it and by official advisers to the program. The problems of the ultimate disposal of radioactive waste remain unsolved. There is an array of other problems.

Mismanagement of the U.S. nuclear power program has become so pervasive that there are now serious questions as to whether the nation can, in fact, manage such a technology with the required care. As a result of this questioning and the consequent public controversy, the U.S. nuclear power industry is now in serious trouble.

Weaknesses in the management of advanced technologies are, of course, by no means new or unheard of. What is distinctive about nuclear power is the uniquely damaging consequences which may stem from such weaknesses. Nuclear technology is perhaps the most unforgiving of all, and requires essentially perfect management to avoid the many possible untoward side effects. Whether the many defects in the present program can be fully and permanently repaired is still unknown.

We believe that the lack of resolution of the problems of nuclear power that have led to the present controversies is unacceptable. The technical and managerial issues must be fully resolved and the construction of new nuclear plants must be restricted until this is accomplished. Moreover, there is need for full public understanding of the special features of this source of energy. It is especially important that there be both an understanding of the potential for harm and an assessment of the likelihood of mischance. Taken together, these two factors determine the social risk. This risk, in turn, can be fairly balanced by the public against the expected gains. This nation might conclude, as did the 1973 Pugwash Conference on Science and World Affairs, that:

"Owing to potentially grave and as yet unresolved problems related to waste management, diversion of fissionable material, and major radioactivity releases arising from accidents, natural disasters, sabotage, or acts of war, the wisdom of a commitment to nuclear fission as a principal energy source for mankind must be seriously questioned at the present time."[5]

The remainder of this volume has as its objective to set forth the technical problems in the nuclear program so as to help move towards that full public understanding which is at present lacking.

References

1 "United States Energy Through the Year 2000," U.S. Department of the Interior (December 1972).

2 "Nuclear Power 1973-2000," USAEC Document WASH-1139, (December 1972).

3 Weinberg, Alvin M. "Social Institutions and Nuclear Energy," *Science*, (July 7, 1972).

4 Ibid.

5 Twenty-third Pugwash Conference on Science and World Affairs, Report of Working Group on Radioactive Pollution of the Environment, August 1973.

Chapter 2

RADIATION-INDUCED LUNG CANCERS AMONG URANIUM MINERS

Arell S. Schurgin, Thomas C. Hollocher

2.1 Summary and Recommendations

Large-scale mining of uranium for atomic weapons began in the southwestern United States in 1946 following the passage of the Atomic Energy Act, and continued through about 1968 when the existence of adequate stockpiles of uranium led to a reduction in mining activities. During this period, and especially from 1946 to 1960, some 6000 underground miners were significantly and needlessly exposed to radioactive gases present in the air of uranium mines.[1]

The health of some 3400 white and 780 nonwhite (American Indian mainly) uranium miners was followed carefully from 1950 to 1968 and less rigorously from 1968 to the present. By 1973[2] an excess of about 180 respiratory malignancies were reported. We estimate that the excess of lung cancer deaths among the total group of some 6000 miners must be currently of the order of 250-300. Predictions of 600-1100 excess lung cancer deaths ultimately due to irradiation have been made for this group.[3] Smoking seems to increase substantially the risk of radiogenic lung cancer, partially at least by shortening the induction period for tumor development. Also, American Indian males seem to show a significantly lower risk than do white males.

The premature death of so many uranium miners must be viewed as a national tragedy and scandal which resulted from the failure of federal authorities to set adequately conservative radiation standards, and to require control programs to achieve and maintain them.[4] The sad fact remains that the present epidemic of lung cancer could have been predicted on the basis of numerous prior investigations in other parts of the world, especially Europe, where radium mining resulted in analogous epidemics of lung cancer. The means for prevention - vigorous ventilation of mines - were known and available, as were the instruments for detecting radiation. The negligence on the part of government and the mining industry was massive and may, in part, represent a deliberate disregard of responsibility. Much of the blame for the epidemic must lie with the AEC, which should have felt the responsibility to inform mine operators and others of the known radiation hazards.

On the basis of our analysis, we offer the following recommendations to help protect the health and lives of present underground uranium miners.

1. The exposure of bronchial epithelium of uranium miners over their working lives (25 years) to irradiation from radon and radon daughters certainly should not exceed 1/10 of the currently estimated doubling dose[5] for the induction of lung cancer in humans, which is taken to be 240 rad.[6] A total dose of 24 rad to the lungs distributed over 25 years of work gives an average annual dose of 1.0 rad/yr (about 6 times background).

Because of the importance of α-radiation, there is a special unit of radiation for radon and its daughters. A "Working Level"

(WL) is defined as any combination of short lived radon daughters in 1 liter of air that will result in the ultimate emission of 1.3×10^5 MeV of potential α-energy.[7] Taking 1 Working Level Month (WLM) to be equivalent to 2 rads to the lungs, the above standard stipulates 0.5 WLM/yr and an average concentration of radon daughters in inhaled air of 0.04 WL throughout the year (i.e., 0.5 WLM/12 months).

This recommendation <u>does</u> <u>not</u> imply a no-damage level. There is no evidence from epidemiological data for a threshold level below which exposure to radon daughters does not lead eventually to some excess <u>risk</u> of lung cancer. Indeed, the dose-response relationship for radiogenic lung tumors <u>as presently understood</u> may underestimate the hazard by a factor of 2 or more. The doubling dose could be as low as 100 rad, and the excess risk may be as high as 2-3 deaths per year per rad per million persons from lung cancer. Irradiation (mainly alpha particles) at a maximum rate of 1.0 rad/yr is suggested because (a) it is only several times the background rate, (b) it provides a lifetime dose that we hope is well below the doubling dose for lung cancer, and (c) it is practically achievable with technology presently available.

This recommendation, 0.5 WLM/yr, may be compared to the present federal standard of 4 WLM/yr or 8 rad/yr, which we and others[8] feel leaves no safety margin and is therefore too high.

2. The power to set and enforce radiation standards in uranium mines should lie within a single federal agency which has no promotional bias toward the use of uranium or nuclear technology. Radiation standards for

uranium miners should also apply to other U.S. miners.

2.2 Radiation and Dosage. Technical Background and Definitions

The natural decay of uranium-238, the principal form of uranium found in mines, leads to the formation of various radioactive products in ore bodies. One of these, radon-222, is a noble gas[9] which can escape from exposed rock surfaces into the air of mine tunnels. As can be seen in Table 2.1, radon-222 has a half life of 3.8 days and decays via alpha (α), beta (β), and gamma (γ) emission through a set of short-lived nuclides (radon daughters) which, as a group, exhibit a half life of about 1/2 hour before reaching lead-210.

Table 2.1 Radioactive decay scheme for Ra-222. The four nuclides, Po-218, Pb-214, Bi-214, and Po-214 are the "radon daughters" of interest in this book.

Nuclide	Half Life	Major Radiation
Ra-222	3.8 day	α
Po-218	3.0 min	α
Pb-214	26.8 min	β, γ
Bi-214	19.7 min	β, γ
Po-214	0.0002 sec	α
Pb-210	20.4 yr	β, γ

Radon daughters, either as free ions or adsorbed to water droplets or dust particles, are inhaled, deposited, and retained in the respiratory system of miners. As a result, respiratory epithelium can be exposed to radiation doses that far exceed whole body doses from internally dissolved radon and from external γ-rays. Lung exposures from radon daughters are far greater than from radon itself, and the bulk of tissue damage is attributed to α-particles, which have short ranges and high linear energy transfer, rather than to β-particles.[10]

Exposure of an individual to radon daughters in the air can be estimated from the length of time the individual breathes an atmosphere containing a stated concentration of radon daughters. A "Working Level Month" (WLM) is defined as exposure to 1 WL of radon daughters for 170 working hours.

Lung cancers believed to arise from radon daughter inhalation appear usually in the hilar region of the lung. This is an area which is ciliated and covered with a moving sheet of mucus. The constantly brushing cilia are bathed in a more watery fluid. The ciliated cells are interspersed with goblet cells that provide the mucous secretion. Since this surface layer of cells develops from a layer of basal cells attached to the basement membrane, the integrity of the epithelium depends on the continued integrity of the basal cells. Thus, the nuclei of these basal cells are taken as the relevant biological target.[11] Estimates of the distance between the biological target, the basal cell nuclei, and the source, α-particles at the mucous layer, are

36 microns[12] minimum and 63 microns median thickness. The range of α-particles from radon daughters (5-7 MeV) in tissue is roughly 39-71 microns. Thus α-particles entering the epithelium will deliver on average a significant fraction of their energies to the basal cell region. With an effective half life of 1/2 hour, it is clear that radon daughters will decay mainly on the bronchial epithelium before they can be removed by normal clearance mechanisms: mucous transport and phagocytosis of foreign particles.

Dust particles larger than about 5 micron are filtered out in the upper respiratory tract and do not reach the deep air passages. Particles in the 0.1-1 micron range and free ions are rather effectively retained. It is apparent, therefore, that actual radiation dosage to the lungs must depend on a number of factors: concentration of radon daughters in the air; respiratory volume and exchange rate; distribution of radon daughters between free ions and ions adsorbed to dust particles; size of dust particles; respiratory clearance rate. This large number of interrelated factors presents a complex picture and requires that estimates of radiation dose to human lungs be based upon explicit physiological models and sets of assumptions which may vary among investigators. For example, the dose in rads from 1 WLM of exposure to bronchial tissue has been calculated to be 2,[7] 2-3,[10,13,14] and 12.[15] We shall take a value of 2 rads per WLM of exposure in this paper.

The rad, as a unit of radiation dosage, measures only the amount of energy deposited per gram of tissue, and does not convey the full biological impact of the exposure. A more relevant measure of the biological effect is the rem (roentgen equivalent man) which takes into account the differential damage caused by different ionizing radiations. The ratio, rem/rad, is equal to 1 for x- and γ-rays, and equal to about 10 for 5-7 MeV α-particles. Thus, a rad of exposure from α-particles does more biological damage than a rad of γ-rays. For the specific irradiation of bronchial epithelium by α-particles from radon daughters, the ratio rem/rad is taken to have a value of three.[8] However, on the basis of empirical epidemiological data on the risk of lung cancer among irradiated humans and animals (to be discussed below), a ratio of rem/rad as low as 1 for miners exposed to radon daughters[8] cannot be ruled out. In the present discussion we shall employ 1 WLM=2 rad = 2-6 rem as applying to the bronchial epithelium of exposed miners.

2.3 Historical Perspective

It has been known since the sixteenth century that many miners who worked for several years in mines in which pitchblende was present developed fatal pulmonary disease. Although first diagnosed as cancer of the lung by Härting and Hesse in 1879,[16] the increased mortality was not generally attributed to lung cancer until the 1920s.[17] The higher incidence of pulmonary disease was recognized particularly in Schneeberg in the Erz Mountains of Germany and in Joachimsthal in Czechoslovakia. Many of the miners of these districts, especially pitchblende (radium) miners, were known to die in middle age of a pulmonary disease which the miners called "Bergkrankheit," or mountain sickness. It occurred chiefly in those miners who had worked in the mines for more than 10 years. The Schneeberg mines employed an average of 650 miners during the years 1869-1877, and the average life expectancy after entering the mines was about 20 years. Hårting and Hesse found that lung cancer accounted for 75% of all deaths and that the incidence was greater in radium miners than in mine masons and carpenters who spent little time below ground. They assumed incorrectly that the cause was inhaled arsenic. It is possible, however, that preexisting respiratory diseases, although not silicosis apparently,[18] and the generally bad state of nutrition were minor predisposing factors in the development of the malignancies.[19]

Several studies between 1900 and 1940[16,18,20-24] established clear statistical evidence of increased rates of lung cancer among these European miners.

The statistical data on lung cancer rates among Schneeberg miners together with the comparative figures for the lung cancer rate in the male population of Vienna, presented by Peller in 1939,[23,25] and given here in Table 2.2, demonstrate the scope of the cancer hazard existing at Schneeberg. Data on Joachimsthal miners are summarized in Table 2.3. The average age at death was 47-49 years[17,26] after an average mining exposure of 17 years (9-23 years).

Estimates of the radioactivity in the air in various mines in the Schneeberg region

Table 2.2 Cancer mortality per 1000 per year in Schneeberg miners.[25]

	Miners in Schneeberg		Vienna Males 15-39 years of age
Cancer of	1875-1894	1895-1912	1932-1936
Lung	12.7 ± 1.0	16.5 ± 1.5	0.34
All other organs	2.4 ± 0.4	2.1 ± 0.6	2.1

Table 2.3 Cancer mortality per 1000 per year in Joachimsthal miners.[25]

	Miners in Joachimsthal		Vienna males 15-39 years of age
Cancer of	1929-1938 Total	1929-1938 Autopsies only	1932-1936
Lung	9.8 ± 1.5	8.3 ± 1.6	0.34 ± 0.01
Larynx	0.2	0.3	0.07
Skin	0.2	0.3	0.05
Bone	0.2	0.3	0.12
All other organs	0.45 ± 0.1	0.0	1.86 ± 0.02
All organs	10.9 ± 1.6	9.2 ± 1.6	2.44 ± 0.03

were first made by Ludwig and Lorenser[24] and later by Rajewsky.[27] The values ranged from 3.6-180 WL. Taking 17 years as the average of work years and 30 WL as the average exposure, the average cumulative exposure was 6120 WLM or more than 12,000 rad! Radiation clearly was the primary cause of "Bergkrankheit." The analogous average exposure for Joachimsthal miners was estimated to be 3840 WLM or about 8000 rad.[22]

In Newfoundland fluorspar mines, where radon entered the mine in ground water, the average concentration of radon daughters in the mine air ranged between 2.5 to 10 WL. The 630 living miners studed by de Villiers and Windish[28] had spent 12 months or more working in the mines over a 29 year period (1933-1961). Of the deaths reported in the single year of study, 26 of 29 were attributed to lung cancer. This incidence is 25-40 times the incidence found in the adult, male population of Newfoundland. The average time spent underground in the fluorspar mines was estimated at 166 months.[29] With concentrations at 2.5-10 WL, the cumulative radiation exposure for these fluorspar miners is estimated to have ranged from 415 to 1660 WLM.

Although numerous studies have consistently shown a high risk of nonmalignant respiratory diseases among underground miners, an excess mortality from lung cancer has not usually been found in the absence of significant exposure to radioactive materials.[30-32] For example, excess lung cancer risk has not been demonstrated among British coal miners exposed to about 0.015 WL, among U.S. potash miners exposed to less than 0.02 WL and among Russian manganese miners exposed to less than 0.3 WL.[7]

2.4 The U.S. Experience

U.S. uranium ores, located primarily in Utah, Arizona, Nevada, Colorado, and New Mexico, were worked for radium (pre-1930), for vanadium (1930-1945), and finally for uranium starting in 1946.[8] The bulk of uranium was mined for weapons production from about 1946 through about 1968. Uranium production subsequently has declined as the result of adequate accumulated stockpiles but is anticipated to undergo major expansion soon to fill the needs of the rapidly expanding commercial nuclear power industry.

Radon measurements were first made in U.S. uranium mines in 1949, but surveys of radon daughter concentrations were not begun until 1951. Until the 1960s, the surveys tended to be spotty although probably representative of uranium mining generally. The purpose through 1960 was apparently to gain information rather than to control mining operations. Major radon daughter control programs were initiated in 1961[8] and continue to the present along with more thorough monitoring. Prior to 1956 it was common for mines to operate at 5-10 WL with a few at 50-200 WL. Advancements in ventilation technology have improved the situation continuously. By 1968 the majority of mines showed values of 0.5-2 WL with about 20% at less than 0.5 WL and none greater than 5 WL.

A possible health hazard to miners was recognized by the Colorado Department of Health in 1949 and has been studied in cooperation with the U.S. Public Health Service

(USPHS). The USPHS published in 1957 the results of radon daughter surveys in U.S. uranium mines and predicted a significant mortality from lung cancer among the miners.[33] Data on radon and radon daughter concentrations were compared with epidemiological and mortality data which eventually encompassed a group of some 4000 uranium miners who had had mining experience from 1950 to 1960.

From a preliminary report in 1962, it appeared that among 907 white uranium miners with more than 3 years underground experience, deaths from respiratory cancer (6 deaths) were occurring at nearly 5 times the normal rate.[34] Though this result was not statistically conclusive it did suggest an impending problem.

Further analysis of the data by Wagoner et al. in 1964,[35] yielded quite strong evidence that an excessive incidence of lung cancer existed among American uranium ore miners. Wagoner et al. found that among 3656 white uranium miners, 218 deaths were observed compared with 148.7 expected ($p < 0.01$). Catagories in which death significantly exceeded that expected were: (a) respiratory neoplasms among uranium miners with 5 or more years underground experience (11 observed as compared with 1.1 expected); (b) "all other causes" - a reflection of pulmonary fibrosis and its complications; and (c) accidents. The 10-fold increase in lung cancer was not attributable to age, smoking, nativity, heredity, urbanization, self-selection, diagnostic accuracy, prior hard rock mining experience, or nonradioactive ore constituents, including silica, arsenic, and dust.

Saccomanno et al.[36] reported that lung

cancer among the miners is seen at an earlier age and with a strikingly higher ratio of undifferentiated, small cell type carcinomas. The frequency of the small cell undifferentiated tumors rises sharply relative to other histological types with increasing radiation exposure to about 1500 WLM then levels off. Correspondingly the relative frequency of epidermoid tumors and tumors of other cell types fall with increasing radiation exposure to about 1500 WLM and then also level off. Thus, undifferentiated carcinomas of the small cell type are so prevalent in uranium miners and in other persons and animals whose lungs have been irradiated (and so relatively rare otherwise), that they represent a signature of heavy radiation exposure.[8]

The first quantitative relationship between the incidence of respiratory cancer and cumulative dosage of radiation was described by Wagoner and his associates in 1965.[37] Using WLMs as cumulative radiation exposure estimates, they reported that the incidence of lung cancer increases from 3.1 per 10,000 miners per year to 116 per 10,000 miners per year over 6 exposure categories ranging from less than 120 WLM to more than 3720 WLM. This relationship persisted even after adjustment for cigarette consumption and other factors. These early data are presented in Figure 2.1. While the confidence limits are wide, the existence of an approximately linear dose-response curve made a casual relationship between airborne radiation and respiratory cancer in U.S. uranium miners virtually certain by 1965.

From 1950 to 1971, an excess of deaths from respiratory cancers has not been statis-

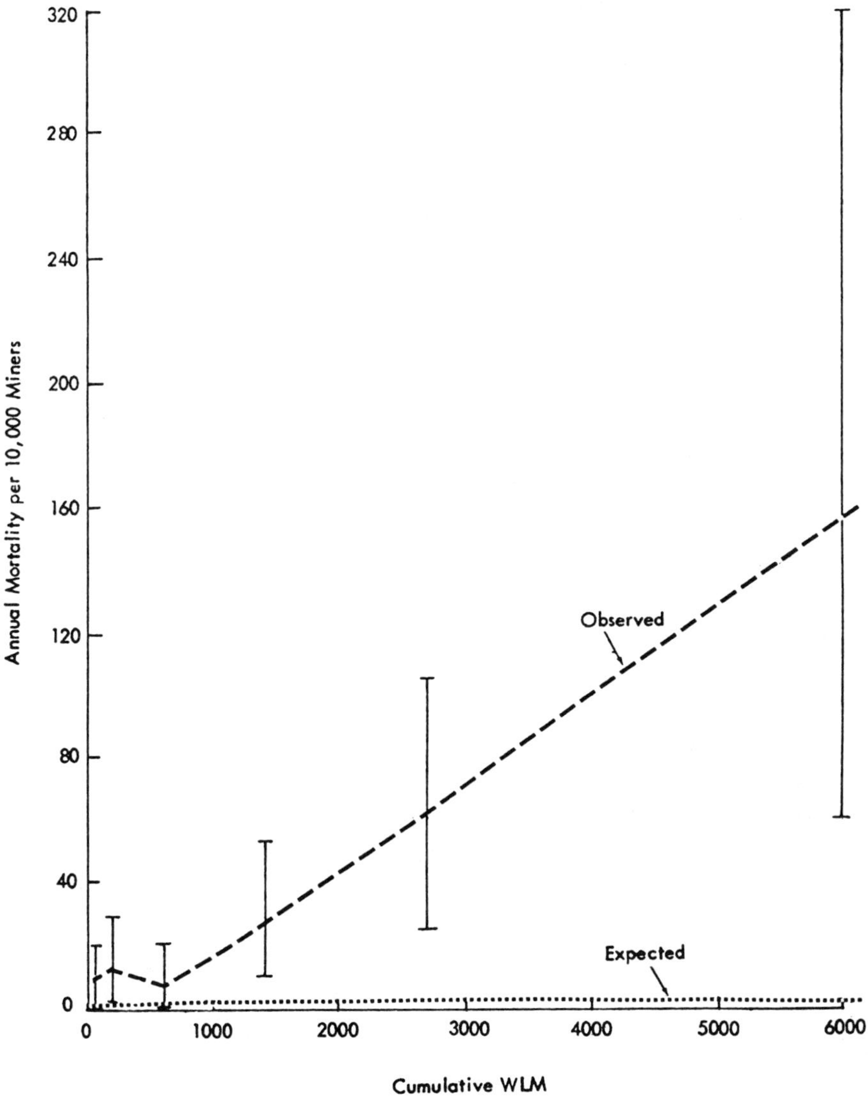

Figure 2.1 Observed and expected annual lung cancer mortality per 10,000 miners (95% confidence limits) against exposure. The expected mortality from lung cancer among adult males in the United States is 2-3 per year per 10,000. From ref. cited in note 1.

tically demonstrated for workers exposed to less than 120 WLM - 4 deaths observed in the study group as opposed to an expectation of 2.2 or, more likely, 3.2 after adjustment for smoking habits. The 1971 study of Lundin et al.[8] confirms a statistically significant excess of respiratory cancer among white uranium miners down to and including the range 120-359 WLM.

The highest incidence of lung cancer is occurring now in the group of miners (a) who worked in mines in which the average concentration of radon daughters was usually greater than 10 WL, (b) whose total cumulative exposures exceeded 1000 WLM, (c) who started mining uranium ore more than 15 years ago, and (d) who were moderate to heavy cigarette smokers.[1]

Although cigarette smoking is associated with an increased risk of lung cancer, the excessive lung cancer among the uranium miners could not be explained by their smoking habits alone. One study published in 1969[38] reported that the ratio of observed mortality from lung cancers among white uranium miners to the expected mortality (based on U.S. white males) was about 4 for both cigarette smokers and nonsmokers. The data are given in Table 2.4. Note that only 2 deaths occurred among the group of nonsmoking miners so that the result lacks statistical significance. Nonsmokers comprised only 25-30% of the miners and their age-specific risk of lung cancer is rather low. A later study[2] gives the mortality ratios as about 6 for nonsmoking miners and 9 for cigarette smoking miners, but again the statistics are poor for nonsmoking miners. Analogous results seem to apply when light

Table 2.4 Respiratory cancer deaths among white uranium miners.

	Person-years	Observed	Expected
Smokers	26,392	60	15.5
Nonsmokers	9,047	2	~0.5

cigarette smokers and heavy cigarette smokers are considered as separate categories. Thus, the results to date suggest that respiratory cancer rates among both smoking and nonsmoking miners are 6-9 times greater than among non-miners with similar smoking habits. There seems to be a roughly multiplicative relationship between smoking and radiation on the rate of incidence of respiratory cancer.

In addition, smoking seems to decrease the induction time required for the onset of radiation-induced lung cancers in uranium miners by 3-4 years, from 20-21 years for nonsmokers (and former smokers who had stopped smoking for more than 8 years) to 17 years for heavy smokers. On the other hand, the distribution of lung cancers among various histological types seems not related to smoking habits. These results, results on the influence of smoking in lung cancers associated with other industrial carcinogens (chromates,[39] asbestos[40]), and animal studies on the effect of cigarette "tar" on the occurrence of skin cancers from β-irradiation[41] all suggest that cigarette smoking may act largely as a promoting agent in the lung cancers of uranium miners.

We see that about 180 excess lung cancer cases were reported by 1973[2] among a group of perhaps 4000 miners (the number was not exactly specified), which included a group of about 3300 white males. Data obtained since 1968 are acknowledged to be incomplete.

Among a total of about 6000 uranium miners over 20 years, the excess incidence of lung cancer is probably 250-300 currently and is expected, according to some,[3] ultimately to reach 600-1100. In addition, exposure to radon daughters is also associated with respiratory symptoms and loss of ventilatory capacity unrelated to overt malignancy, but including metaplastic and probably precancerous changes.[42,43]

2.5 Epidemiology of Lung Cancer Induced by Radiation

The lowest estimated radiation doses associated with excess lung cancer rates in human populations are: 15 rad for atom bomb survivors; 77 rad for spondylitis patients who have undergone X-ray therapy for this rheumatoid condition; 116 rad for Thorotrast patients[44]; 520 rad for British hematite miners; and 360 rad (possibly as low as 240 rad) for uranium miners.[8] Among animals, lung cancers are known to result from 71 rad to rats from inhaled polonium-210 and from 100 rad of X-rays to dogs and mice. The normal incidence of lung cancer among all U.S. adult males is about 2 per ten thousand per year. The doubling dose among U.S. uranium mine workers has been estimated to be as low as 120 WLM or 240 rad (if 1 WLM = 2 rad).[8,29] Table 2.5 summarizes the available data on the increased incidence of lung cancer among irradiated humans.

The epidemiological summary of Table 2.5 does not exclude the possibility that the assumed conversion 1 rad = 3 rem for inhaled radon daughters may actually be as low as 1 rad = 1 rem.

It is important to note that the analyses on U.S. uranium miners may underestimate the actual risk, which may be as much as twice that now indicated because: (a) The sampling of air by mine inspectors is known to be biased toward areas with high WL values. (b) Studies of other groups - notably spondylitis patients and atomic bomb survivors - suggest that excess lung cancer occurs at lower (less than 100 rem) exposure levels than could be studied adequately up to the present among uranium miners. (c) Experimental studies with animals confirm that lung cancer is quite radiogenic and that cancer may result from 100 rad or less. (d) The study group has been compared in the literature to U.S. males and this may not be appropriate. In fact, the miners come largely from nonurban areas of the Southwest where the normal incidence of lung cancer among males is only about 2/3 that of the United States generally. (e) The risk of respiratory cancer per unit exposure appeared to be greater in the lower cumulative radiation groups than in the higher ones, as of 1971. Thus the dose-response curve may be worse than linear and show increasing slope at lower doses. (f) Assignments of cumulative-dose favored high-dose estimates in cases of ambiguity or doubt.[25] (g) The latency period at low doses may actually exceed the 10-20 years that is now assumed to apply.

Table 2.5 Excess respiratory cancer in irradiated humans.[7]

Group at risk	Excess respiratory cancer cases per year per 10^4 persons		
	per 100 rad	per 100 rem	Radiation
U.S. uranium miners	0.9	0.3-0.9[a]	Inhaled radon daughters
Joachimsthal and Schneeberg miners	1.3	0.4-1.3[a]	"
Newfoundland fluorspar miners	1.1	0.3-1.1	"
British hematite miners	3.0	1.0-3.0[a]	"[b]
U.S. hard rock miners	1.0	0.3-1.0[a]	"
Thorotrast patients (Portuguese)	1.4	0.5[c]	Internal radon daughters

Table 2.5 Excess respiratory cancer in irradiated humans (continued).

Group at risk	Excess respiratory cancer cases per year per 10^4 persons per 100 rad	per 100 rem	Radiation
Thorotrast patients (Danish)	2.0	0.7[c]	"
Spondylitis patients	3.3	3.3[d]	X-rays
Atomic bomb survivors	4.0	2.0	γ-rays and neutrons

a. rem/rad taken to be 3 for α-emitters on bronchial epithelium, but comparison of epidemiological data from spondylitis patients and atomic bomb survivors suggests the ratio may be as low as 1.

b. Hematite may be a promoter or co-carcinogen.

c. rem/rad taken to be 3

d. Possibly too high as the result of mistaking lung cancer for spondylitis.

In summary, it has been found that some 4000 U.S. uranium miners, who comprised a study group, received from radon daughters a median occupational exposure of 700-1000 rad to their bronchial epithelium from 1946-1960. Maximum exposures are believed to have been as high as 10,000 rad. The average exposures became gradually lower during the 1960s as radon control programs became widespread. The results to date suggest a doubling dose for lung cancer of about 240 rad for white males after a latency period of some 10-20 years. This implies an excess rate of fatalities from lung cancer of about 1 per year per rad per million persons among white males.

2.6 Radiation Standards

The present radiation protection guide in uranium mines is 4 WLM/year.[47] This corresponds to 200 rad over a 25-year work period and probably approaches the doubling dose for lung cancer. The standard has been criticized by Lundin et al.[8] as having no safety factor and thus being excessively high. While there is bound to be considerable disagreement as to an appropriate standard, we suggest that cumulative lifetime (25 year) exposure in mining should not exceed 1/10 of that dose which is believed to induce lung cancer in man or animals (about 100 rad) and certainly should not exceed 1/10 of the apparent doubling dose (about 240 rad). We suggest, therefore, that the cumulative dose to lungs should not exceed 25 rad at the rate of 1.0 rad/yr (about 6 times background exposure). This corresponds to about 0.5 WLM/yr or a concentration of 0.04 WL.[49] There is no evidence at present that a threshold exists for the induction of lung cancers by α-radiation in humans. If a threshold does exist, it probably lies below the level of 50 rad.

2.7 Feasibility of Standards

Radon and its daughters can be removed from mines by ventilation. Present technology permits many uranium mines to operate at less than 0.5 WL and 0.2 WL is commonly achieved in better mines. Our recommendation of 0.04 WL would require substantial improvements in ventilation. Alternatives involve: providing miners with separate air supplies and/or air filters; the undertaking of underground mining by remote controlled machines; or mining all uranium from open pit mines. In any case, the additional cost should be considered a normal cost of operating a safe fission technology.

The demand for uranium in the nuclear power industry has been estimated to be about 1/4-1/3 million tons from the present through 1980-1985, at a total value of $5-10 billion. Very limited data on a group of 14 mines[1,4] suggest that the average total (investment plus 10 years' operation) costs of ventilation and radon control systems to achieve 1 WL are $0.3-1.5 million per mine. Since about 500 underground mines were producing uranium in the United States from 1961 to 1966 and 850 were producing during the peak years 1957-1958, we will assume that something on the order of 500 underground mines will contribute to the production of uranium for commercial power. We also assume that the average additional cost per mine over 10 years will be $2 million to achieve 0.04 WL. The additional cost comes to about $1 billion or

10-20% of the value of the uranium mined.

2.8 Failure of the AEC to Protect Miners

The matter of radiation protection in the mines was one of confusion until 1967 when uniform FRC standards came into effect. In addition to the several state authorities responsible for mining practices and for health and safety standards, numerous federal agencies have been involved, including the Department of Interior under the Federal Metal and Nonmetallic Mine Safety Act, and Health, Education, and Welfare in providing technical advice on matters of health standards and control of hazards.[1] As a result partly of divided authority, a possible health hazard from radon daughters was not recognized until about 1949; methodical air monitoring programs were not in operation until about 1954; and radon control programs were not very effective or extensive before 1961. Currently the regulation of uranium mining practices falls largely under state authority.

The AEC failed to regulate uranium mine practices and to protect the health of miners because of a peculiar interpretation of the Atomic Energy Act. The Act was interpreted _by the AEC itself_ not to include uranium mining but, at the same time, to apply to all other aspects of uranium processing and use.[4] The view was held that the AEC could regulate source material (i.e., ores of uranium and thorium) only after removal from "its place of deposit in nature." But, the AEC considered the _mine_ rather than the _ore body_ as the "place of deposit in nature." This narrow, self-serving interpretation was considered too restrictive even by the Joint Committee on Atomic Energy, at least in 1959.[51] To our knowledge, the interpretation was never challenged in court.

Metzger[52] has analyzed the legal basis of the AEC's interpretation regarding the regulation of uranium mining. He points out that the Atomic Energy Act is very broad in scope and provides in fact for the regulation of uranium ore under all circumstances. A license is required even for possession of uranium ore. Furthermore, the Act nowhere excluded the licensing of mines. Even failing authorization under the Atomic Energy Act, the AEC could have regulated mining and required safe standards under the terms of the Walsh-Healy Public Contracts Act of 1936, which provides for federal supervision of worker safety in situations involving federal contracts. All AEC contractors fall under the Walsh-Healy Act and, accordingly, the AEC had direct control over their procedures and operations. A simple directive to mine owners would have resulted in the necessary changes having been made, with the costs being borne by the AEC.

Notes and References

1 Federal Radiation Council, Guidance for the Control of Radiation Hazards in Uranium Mining, Report No. 8, revised, U.S. Government Printing Office, Washington, D.C., 1967.

2 V. E. Archer, J. K. Wagoner, and F.E. Lundin, J. Occupational Med. 15 204 (1973); Health Phys. 25 351 (1973).

3 C. C. Johnson of the U.S.P.H.S. in a speech at a conference of the National Tuberculosis and Respiratory Disease Association, 1969.

4 Joint Committee on Atomic Energy, Subcommittee on Research, Development, and Radiation. Congress of the United States, 90th Congress, May-August 1967, Hearings on Radiation Exposure of Uranium Miners, Parts 1 and 2, U.S. Government Printing Office, Washington, D.C., 1967.

5 A "doubling dose" for a given form of cancer is that radiation dose which will lead to a doubling of the spontaneously occurring rate of that cancer.

6 A rad (Radiation Absorbed Dose) of radiation exposure is equal to 100 ergs of energy deposited in one gram of tissue.

7 D.A. Holaday, D.E. Rushing, R.D. Coleman, D.F. Woolrich, H.L. Kusnetz, and W.F. Bale, Control of radon and daughters in uranium mines and calculations on biologic effects, PHS Publication No. 494, U.S. Government Printing Office, Washington, D.C., 1957.

8 F.E. Lundin, J.K. Wagoner, and V.E. Archer, _Radon Daughter Exposure and Respiratory Cancer, Quantitative and Temporal Aspects_, National Institute for Occupational Safety and Health and National Institute of Environmental Health Sciences Joint Monograph No. 1, National Technical Information Service, Springfield, Virginia, 1971.

9 Noble gases, such as helium, neon, and argon, are those that are inert chemically.

10 W.F. Bale and J.V. Shapiro, _Proceedings U.N. International Conference on Peaceful Uses of Atomic Energy_, August 1955, Vol. 13: pp. 233-236, United Nations, New York, 1956.

11 H.M. Parker, _Health Phys_. _16_ 533 (1969).

12 One micron equals 1 millionth of a meter.

13 B. Altschuler, N. Nelson, and M. Kuschner, _Health Phys_. _10_ 1137 (1964).

14 W. Jacobi, _Health Phys_. _10_ 1163 (1964).

15 A.K.M.M. Haque, and A.J.L. Collinson, _Health Phys_. _13_ 431 (1967).

16 F.H. Härting, and W. Hesse, _Viertiljahrsschr. gerichtl. Med. u. offentl. Gesundheitswesen_ _30_ 296; _31_ 102, 313 (1879). Cited by Lorenz, 1944.

17 H. Sikl, _Acta Unio Internationalis Contra Cancrum_ _6_ 1366 (1950).

18 W. Hueck, _Ztschr. Krebsforsch_. _49_ 312(1939).

19 E. Lorenz, J. Nat. Cancer Inst. 5 1 (1944).

20 O. Rostoski, E. Saupe, and G. Schmorl, Ztschr. Krebsforsch. 23 360 (1926).

21 H. Sikl, Ztschr. Krebsforsch. 32 609 (1930).

22 A. Pirchan and H. Sikl, Amer. J. Cancer 16 681 (1932).

23 S. Peller, Human Biol. 11 130 (1939).

24 P. Ludwig and E. Lorenser, Ztschr. Phys. 22 178 (1924); Strahlentherapie 17 428 (1924).

25 W.C. Hueper, Occupational and Environmental Cancer of the Respiratory System, Charles C. Thomas, Springfield, Illinois, 1942.

26 A. Glucksmann, L.F. Lamerton, and W.V. Mayneord. In Ronald W. Raven ed., Cancer, Volume I, Butterworth and Co., London, 1957.

27 B. Rajewsky, Ztschr. Krebsforsch. 49 315 (1939).

28 A.J. deVilliers, and J.P. Windish, British J. Indust. Med. 21 4 (1964).

29 V.E. Archer, and F.E. Lundin, Environmental Res. 1 370 (1967).

30 J.K. Wagoner, R.W. Miller, F.E. Lundin, J.F. Fraumeni, and M.E. Haig, New Eng. J. Med. 269 284 (1963).

31 C.B. Chatgidakis, Arch. Environ. Health 1 17 (1960).

32 R. Paul, Arch. Environ. Health 2 96 1961).

33 H.N. Doyle, Radiation Hazards in Uranium Mining, Occupational Health Program, Div. of Special Health Services, USPHS, Washington, D.C., 1957.

34 V.E. Archer, H.J. Manguson, D.A. Holaday, and P.A. Lawrence, J. Occupational Med. 4 55 (1962).

35 J.K. Wagoner, V.E. Archer, B.E. Carroll, D.A. Holaday, and P.A. Lawrence, J. Nat. Cancer Inst. 32 787 (1964).

36 G. Saccomanno, V.E. Archer, O. Auerbach, M. Kuschner, R.P. Saunders, and M.G. Klein, Cancer 27 515 (1971).

37 J.K. Wagoner, V.E. Archer, F.E. Lundin, D.A. Holaday, and J.W. Lloyd, New Eng. J. Med. 273 181 (1965).

38 F.E. Lundin, J.W. Lloyd, E.M. Smith, V.E. Archer, and D.A. Holaday, Health Phys. 16 571 (1969).

39 H. Oettel, A.M. Thiess, and C. Whl, Zbl. Arbeitsmed. 18 291 (1968).

40 I. Selikoff, E.C. Hammond, and J. Churg, J. Amer. Med. Assoc. 204 106 (1968).

41 E.V. Cowdry, A. Croniger, S. Solerie, et al., *Cancer* *14* 344 (1961).

42 V.E. Archer, H.P. Brenton, and J.K. Wagoner, *Health Phys*. *10* 1183 (1964).

43 V.E. Archer, B.E. Carroll, and H.P. Brinton, *Radiological Health and Safety in Mining and Milling of Nuclear Materials*, Vol. 1, pp. 21-36, International Atomic Energy Agency, Vienna, 1964.

44 Thorium dioxide in a preparation called Thorotrast was widely used as an X-ray contrast medium during the 1930s. Thorotrast thus provides an internal source of α-radiation. A typical single intravenous dose of Thorotrast has been calculated to deliver about 13 rad/yr to lung (whole lung, not just epithelium). An excess of leukemia and of lung and liver tumors have been reported among two groups of patients who have received Thorotrast. (See notes 45, 46.)

45 J. Horta, and L.C. deMotta, *Ann. New York Acad. Sci*. *145* 830 (1967).

46 M. Faber, *Ann. New York Acad. Sci*. *145* 843 (1967).

47 The present standards for uranium miners, effective July 1, 1971, (see note 48), stipulate that no individual will receive an exposure of more than 2 WLM in any consecutive 3-month period and no more than 4 WLM in any consecutive 12-month period (see notes 1, 48). From 1967 to 1971 the yearly standard was 12 WLM. Federal standards were not in effect

prior to 1967, despite the recommendation of 1 WL made by the PHS ten years earlier. The International Commission on Radiological Protection, the National Council on Radiation Protection and Measurements, and the Federal Radiation Council (FRC) have all recommended 15 rem/yr as the maximum permissible dose-equivalent to lungs for occupational workers since about 1960 (see note 4). This is an ambiguous standard, since the conversion from rem to WL is not precisely known. Roughly, 15 rem/yr = 2.5 WLM. The present FRC-EPA standard, therefore, is roughly consistent with the recommendations for occupational workers. The occupational standard of 5 rem/yr whole-body dose (as from highly penetrating γ-rays) should not be confused with the standard of 15 rem/yr, which applies to individual organs of the body.

48 Federal Register 33, No. 252 (1969); 34, No. 10 (1969); 35, No. 245 (1970); 36, No. 101 (1971).

49 It should be noted that γ-rays in uranium mines deliver a whole-body dose (including lungs) in the range of 1-2 mrad/hr (see note 1). Over a 2000-hr work year, 1 mrad/hr would lead to an accumulated dose of 2 rad (2 rem) to lungs and other organs. Over 25 years this amounts to 50 rad, which is roughly the doubling dose in humans of the more radiogenic cancers, such as leukemia (see note 50), and about 1/5 the doubling dose for lung cancer. The γ-ray dose occurs in addition to any dose to lungs received from radon daughters in air.

50 BEIR Report: The Effects on Populations of Exposure to Low Levels of Ionizing Radiation, Div. of Medical Sciences, NAS-NRC, Washington, D.C., 1972.

51 Employee Radiation Hazards and Workmen's Compensation, summary analysis of hearings, Joint Committee on Atomic Energy, U.S. Congress, Washington, D.C., March 1959.

52 H.P. Metzger, The Atomic Establishment, Simon and Schuster, New York, 1972.

Chapter 3

RADIATION HAZARDS ASSOCIATED WITH URANIUM MILL OPERATIONS

Thomas C. Hollocher, James J. MacKenzie

3.1 Summary and Recommendations

The extraction of uranium from ore results in a sand-like material, called tailing, which still contains most of the original radium, and in some fluids and slimes which also contain radium. Prior to 1959 the disposal and handling of these materials caused the pollution of streams and water supplies by radium in the southwestern United States. Prior to 1966 uranium tailings were used as fill and for other construction purposes around buildings. Some 3300 buildings in and around Grand Junction (Mesa County), Colorado, have been discovered to have had tailings used as fill under or against foundations or as a component in construction materials. The total is some 5000 buildings in the whole of Colorado and perhaps another 1000-2000 buildings elsewhere in the United States. One can increase these figures by 50-100% by considering, in addition, the total number of sites away from buildings where tailings have been used as fill and where contamination is due to the erosion of tailings piles or the spillage of uranium ore itself.

The tailings emit gamma (γ) rays and radon gas. When tailings are used in building construction, this radiation and gas reaches persons indoors. Extensive surveys have been carried out in Grand Junction from 1966 to the present (1973) in an effort to quantify radiation levels in buildings associated with tailings. A common γ-ray level indoors due to tailings is about 10 μrad/hr[1] above the normal γ-ray background and the highest levels are in the range 100-300 μrad/hr. Dose rates at these levels would range between 80 mrad/yr and 0.8-2.4 rad/yr whole body dose, respectively. Eighty mrad/yr is about half of the total background dose in Grand Junction and may be considered a relatively low dose. On the other hand, 0.8-2.4 rad/yr is well above two times background and should be considered a significant dose in the general population.[2]

The median integrated levels of radon daughters in air indoors due to tailings in the 600 buildings studied is about 0.016 WL[3] above the normal radon daughter background of about 0.005 WL. This could result in a maximum dose to lungs of about 2 rad/yr. The highest integrated levels were in the range of 0.4-1.5 WL. This could result in a maximum dose to lungs of 40-150 rad/yr. Both the median and highest levels could deliver significant doses to lungs and require remedial action. The doubling dose for radiation-induced lung cancer is about 240 accumulated rad, and the induction rate is about 1 case per year per rad per million persons.

The problems of radiation exposures from tailings arose because the AEC failed to regulate properly the use and disposal of tailings and, in fact, attempted to absolve itself of responsibility through narrow interpretations of the Atomic Energy Act.

In view of this situation and in anticipation of the milling of uranium ore which is soon to begin for the nuclear power industry, we make the following recommendations.

1) With regard to external γ-ray doses received from mill tailings already used in building construction, remedial action (such as removal of tailings or relocation of structures) should be based on the Surgeon General's guidelines of 1970: No action indicated at less than 50 μrad/hr; action suggested, though not imperative, at 50-100 μrad/hr; some remedial action definitely to be taken at greater than 100 μrad/hr. Action is thus suggested for only about 5% of the structures affected. The figures refer to dose rates above background.

2) With regard to lung doses received from radon daughters originating from radon diffusing from tailings, remedial action (such as removal of tailings, improved ventilation, and vapor barriers) should be based on guidelines somewhat more stringent than those of the Surgeon General, namely, those of Table 3.1. Action is thus suggested or indicated in about half of the structures affected (perhaps 3000 structures throughout the United States).

3) Present piles of tailings, amounting to some one hundred million tons, and tailings from future operations should be disposed of by burial either in old uranium mines or as part of land restoration projects associated with strip or open pit mining. Burial should be to a depth of at least 5 meters and should involve as great a dilution of tailings with other mineral material as is practicable.

We consider the AEC proposal[4] to maintain tailing retention ponds over some 20 years for liquid and solid mill wastes to be a totally unsatisfactory method of permanent dis-

Table 3.1 Guidelines for taking remedial action in the case of long-term exposure of the general population to lung doses from the inhalation of radon daughters.

	WL above background[a]	
	Surgeon general, 1970	Suggested by UCS
No action recommended	less than 0.01[b]	less than 0.01
Action suggested	0.01-0.05	0.01-0.025
Action indicated	greater than 0.05	greater than 0.025

a. Constant exposure to 1 WL of radon daughters would deliver a dose to lungs of about 100 rad/yr. One WL is defined as any combination of short-lived radon daughters in 1 liter of air that will result in the ultimate emission of 1.3×10^5 MeV of potential alpha energy.

b. Background is about 0.005 WL indoors in Grand Junction, Colorado. In most other parts of the United States it is less than 0.005 WL.

posal. See Table 3.2 for the decay series of U-238.

4) A study of more sophisticated methods to decrease the radium hazard from tailings should be undertaken. It would appear, for example, that the chemical extraction and purification of radium at the time of the extraction of uranium is possible and that the resulting radium could be incorporated into reactor fuel under an ongoing actinide burnup program and destroyed through transmutation and fission in reactors. Thermal neutron reactors would appear to be adequate for this purpose.

3.2 Production of Mill Tailings

Radiation exposure from uranium ore is not limited to miners[5] and includes parts of the general population as well as the ecology of streams adjacent to uranium mills.

After mining, the ore undergoes an in-plant process known as milling. The ore is finely ground and most of the uranium and thorium is chemically extracted. The resulting spent ore, resembling fine sand and called tailings, still contains, however, most of its original radium - typically 700 pCi/g[6] which is more than 100 times that found in ordinary rocks. In addition, radium is also found in liquid wastes from the extraction process and in a mud-like by-product referred to as slime. Slimes and liquid wastes were usually disposed of in nearby streams after dilution to appropriate radiation levels for water. Most tailings were accumulated in some 35 large piles on mill sites around the mining district of the western United States, primarily on the Colorado plateau, and amounted to about one hundred million tons of

Table 3.2 The Decay Series of U-238.[a]

Nuclide	Half-life	Major Radiation
U-238	4.5×10^9 yr	α
Th-234	24.1 day	β, γ
Pa-234	1.2 min	β, γ
U-234	2.5×10^5 yr	β, γ
Th-230	8.0×10^4 yr	α, γ
Ra-226	1630 yr	α, γ
Rn-222	3.8 days	α
Po-218[b]	3.0 min	α
Pb-214	27 min	β, γ
Bi-214	20 min	β, γ
Po-214	1.6×10^{-4} sec	α
Pb-210	20.4 yr	β, γ
Bi-210	5.0 day	β
Pb-206	Stable	

a. Ore with 0.25% uranium will contain about 700 pCi of Ra-226 per g. At decay equilibrium, which would apply in the ore, the activity of all radioactive species would be equal, e.g., 700 pCi/g. Beyond Ra-226, only Pb-210 has a long half-life.

b. The short-lived radon daughters which cause most radiation exposure to lungs are Po-218 through Po-214.

material from 1946 to 1968.

3.3 Radium in Water Supplies

Because of the small average particle size in tailings, radium is easily dissolved out by water, radon gas easily diffuses away, and the piles of material themselves are subject to erosion by wind and water.

It was finally recognized in the late 1950s by state health authorities, with the assistance of the Federal Water Pollution Control Agency, that a number of streams had become badly polluted with radium and that water supplies for both drinking and irrigation were endangered.[7,8] For example, in the Animas River (downstream from the AEC-licensed uranium mill at Durango, Colorado) damage to fauna attributable to radium could be traced 50 miles downstream by 1958. Radioactivity, principally radium, was 500 times the background in the stream and 8 times Radiation Protection Guide (RPG) limits.[9] About 30,000 people lived along the Animas below Durango, primarily in Aztec and Farmington (N.M.) and they used the river exclusively for drinking and irrigation. Drinking water in these towns frequently contained radioactivity exceeding RPG limits.[10] Due to sequestration of minerals in food chains in the streams, flora and fauna were found to concentrate radium 10^2-10^4 times that found in the water itself. The resulting internal levels were often toxic or lethal to flora and fauna. Grasses and alfalfa irrigated with the river water concentrated radium on the order of 100 fold, and this was then passed on to livestock.

Similar and often higher concentrations of radium were found downstream from uranium mills elsewhere in Colorado. Radium bearing sediments even migrated into the Colorado River, which supplies water for much of the Southwest, including the populous Los Angeles region. In 1959 mills finally were forced by the U.S. Public Health Service and by new AEC regulations to stop or diminish their dumping of radioactive effluent into the streams. By the 1960s, radium contamination of the Colorado River basin had diminished considerably. However, most piles of tailings remained uncovered or otherwise unstabilized and continued to be eroded by wind and water.

3.4 Radiation Hazards from Mill Tailings

It was discovered by the AEC in 1958 that radon emanating from the piles provided rather high level of airborne radioactivity in their immediate vicinity.[11] This plus significant γ-ray levels at their surfaces suggested that the piles ought to be classified as restricted areas.[12] The public was not excluded from access to the piles, however, until 1969.[10,13]

More significant to the general population, however, is exposure from γ-rays and radon daughters[14] in houses and other buildings built upon mill tailings or around which tailings had been used as fill material.[8,10,15] In addition, tailings apparently have been used as a substitute for sand in concrete and mortar. The practice was discovered in 1966 by Colorado Public Health officials in Grand Junction, which contains the large accumulated tailings of the Climax Uranium Company.[8,10] The officials actually observed tailings being used as fill on a construction project, and they subsequently noticed what appeared to be anomalously high readings for γ-radiation and for radon daughters in certain buildings. Investigation revealed that 2-3 x 10^5 tons of tailings had been used in construction in Grand Junction for at least 15 years prior to 1966 and that about 3300 homes and buildings have tailings under or around their foundations. The tailings had been removed from the mill and used with the knowledge and permission of the AEC. Central records were not kept of the removals, so the amounts and location of tailings were determined largely from γ-ray surveys. In Colorado alone some 5,000 buildings had tailings used for construction purposes and there may be a total of a couple of thousand more elsewhere in the West. These figures must be increased by 1.5-2 fold if one includes other sites away from buildings where contamination has been noted. These include sites where mill tailings have been used as land or road fill, where mill tailings have been transported by wind and water, and where uranium ore itself has been spilled or piled up.[10]

That the AEC was aware as early as 1961 of a possible health hazard from the use of mill tailings in construction is indicated by the now famous "1961 letter" (see Appendix 3.1) which was prepared by Harold L. Price of the AEC and allegedly sent to the Public Health Departments of nine states where uranium was mined or milled: Colorado, Utah, Wyoming, New Mexico, Arizona, Oregon, Washington, South Dakota, and Texas. None of these agencies seem, however, to have the letter in their files,[10] and all but one admit complete ignorance of the letter.[15]

The use of tailings in construction leads to radiological hazards in two ways. The radium decay scheme gives rise to γ-rays which can penetrate slabs and masonry walls and cause whole-body exposure inside. Radon is known now to diffuse through concrete and other porous materials and to cause an air-borne source of radiation indoors just as in uranium mines.

The levels of radiation detected in the affected buildings are not very high generally, but they are significant and constitute a possible radiological hazard. While the γ-ray levels are easily mapped, the levels of radon and radon daughters in the air are most difficult to determine precisely due to large differences in the rate of ventilation of buildings as a function of weather, season, and wind speed. In general, a building must be surveyed for a year with numerous readings to arrive at some appropriately weighted average or integrated level of radon daughters. Long term air sampling had provided integrated radon daughter levels for only 47 buildings as of late 1971 (Table 3.3). The latest survey (Table 3.4) extends the data to some 600 buildings, most of which are homes.[16,17]

Roughly 20% of the 47 integrated radon daughter values available in 1971 were reported to be above 0.1 WL (background, indoors, in Grand Junction was reported in 1971 to be about 0.004 WL).[18] The median integrated values were 0.02-0.03 WL. Eighty-seven percent of structures examined which showed γ-ray levels greater than 20 μrad/hr also showed integrated values for radon daughters greater

Table 3.3 Integrated air sample results from buildings in Grand Junction, Colorado, 1971.[a]

WL range[b]	Frequency	Rel. Frequency (%)
Control Locations:		
< 0.005	9	64
0.005-0.013	5	36
≥ 0.014	0	0
	14	100
Tailings Locations:		
< 0.005	1	2
0.005-0.013	5	11
0.014-0.053	26	55
≥ 0.054	15	32
		100

a. Data from ref. cited in note 26. See text for a discussion of the errors of this study.

b. Includes background estimated at 0.004 WL.

than 0.013 WL. Individual readings up to 1.88 WL were found in structures where tailings had been used and up to 0.007 WL where tailings had not been used.

The 1971 integrated air sampling results were suspect for two reasons. First, an instrument artifact gave readings erroneously low by a factor of about 2 (25% low in the case of background). Thus, the median integrated values were actually about 0.05 WL. Second, the sample of buildings studied was not representative of all buildings associated with tailings. It was biased toward commercial structures rather than homes and toward structures that were expected to give high air levels on the basis of high observed γ-ray levels. As is now known, however, γ-ray levels are a poor index of integrated levels of radon daughters. It is now believed that these two kinds of opposing bias cancelled each other more or less.[16] The extended survey, completed in 1973 on some 600 buildings in a representative sample, exhibits a median value for integrated air samples of about 0.021 WL with extreme values in excess of 0.5 WL. The radon-daughter background in the Grand Junction area is now taken to be 0.005 WL indoors.

Table 3.5 summarizes the γ-ray data as of late 1973 over some 5000 indoor sites in Colorado. The median γ-ray level is about 13 μrad/hr and background is taken to be 10 μrad/hr. The γ-ray background of 10 μrad/hr over an 8000 hour year yields 80 mrad/yr whole-body dose out of a total background in Colorado of perhaps 150 mrad/yr. The median <u>additional</u> dose due to tailings is about 3 μrad/hr and represents maximally (i.e., constant exposure) only about 24 mrad/yr. About 20% of the sites exhibit values in excess of 20 μrad/hr. In those 5% of the sites where the additional exposure from tailings is greater than 50 μrad/hr, the average additional dose becomes about 450 mrad/yr or about 3 times total background. The Surgeon General's guidelines of 1970 recommend remedial action at greater than 100 μrad/hr above background for γ-rays and suggests remedial action from 50-100 μrad/hr above background.

The hazard from radon daughters is perhaps greater than that from γ-rays. If we consider an integrated value above background of 0.025 WL over 1 year, continuous exposure would involve about 50 working month units of 170 hours each per year for a resident. This amounts to 1.25 WLM accumulation/yr, and to a lung dose of about 2.5 rad/yr. Over 30 years this amounts to nearly 75 rad to the lungs. This is about 1/3 of the doubling dose for lung cancer and potentially involves several thousand persons. Levels in about 5% of the buildings associated with tailings would be greater than 10 times this dose if the data of Table 3.4 are indeed representative.

As discussed in the previous chapter, the occupational standard for radon daughters in uranium mines is 4 WLM/yr, which could be accumulated by working in air containing 0.3 WL of radioactivity above background for 12 working months. As a general rule, occupational dose standards tend to be about 10 times higher than those for the general population.[2] Thus one might expect a limit of

Table 3.4 Integrated indoor radon survey for buildings associated with mill tailings in Grand Junction, Colorado.[a]

WL range[b]	Number of buildings	Percent of buildings	Approx. maximum dose to lungs (rad/yr)
0-0.005	62	10	0-0.5
0.005-0.01	133	21.5	0.5-1
0.01-0.025	170	27.5	1-2.5
0.025-0.05	118	19	2.5-5
0.05-0.1	74	12	5-10
0.1-0.2	31	5	10-20
0.2-0.4	18[c]	3	20-40
0.4-0.6	5[c]	0.8	40-60
0.6-0.8	2[c]	0.3	60-80
0.8-1	3[c]	0.5	80-100
1-1.5	2	0.3	100-150
1.5-2	1		150-200

a. Survey of 1971-72 on 619 buildings, ref. cited in note 16, EPA computer printout of January 21, 1974.

b. Includes background estimated at 0.005 WL. 1 WL of radon daughters delivers about 100 rad/yr to lungs under constant exposure.

Table 3.4 Integrated indoor radon survey for buildings associated with mill tailings in Grand Junction, Colorado (continued).

Median level, 0.021 WL; average, 0.057 WL. Integrated values are generally from data over 6 locations within each structure.

c. Values at individual locations frequently were in the range of 0.5-1.5 WL.

Table 3.5 Indoor γ-ray survey for buildings associated with mill tailings in Colorado.[a]

Scintillation counter readings, S	Corrected γ-ray levels[b] (μrad/hr)	Number of buildings	Percent of buildings
0-25	0-14.9	3766	69
24-40	14.9-23.2	715	13
40-60	23.2-34.5	342	6
60-100	34.5-57	272	5
100-200	57-113	283	5
200-300	113-169	41	0.8
300-500	169-281	24	0.4
500-1000	281-561	4	0.1
> 1000	> 561	1[c]	

a. Survey of 5448 buildings through late 1973, ref. cited in note 16, EPA computer printout of January 22, 1974: Colorado Long Form HIG (Highest Indoor Gamma) Listing.

b. Corrected γ-ray level = 0.56S + 0.9 in μrad/hr and includes γ-ray background estimated at 10 μrad/hr. Median level, 13 μrad/hr; average 23 μrad/hr.

Table 3.5 Indoor γ-ray survey for buildings associated with mill tailings in Colorado (continued).

c. Factory.

0.4 WLM/year = 0.8 rad ~ 0.01 WL above background for a resident. The Surgeon General's Guidelines of 1970 are consistent with this view and recommend action at greater than 0.05 WL and suggest action for 0.01-0.05 WL.

We of the Union of Concerned Scientists believe that constant exposure to even 0.025 WL is excessive for the general population and recommend guidelines for remedial action in Table 3.1 somewhat more stringent than those of the Surgeon General. Namely, we would suggest remedial action in the range 0.01-0.025 WL and require action at greater than 0.025 WL. From information presently available as the result of integrated air sampling, one may expect something of the order of 3000 buildings contaminated with mill tailings over the United States to receive remedial action. This constitutes about half of the total buildings so contaminated.

3.5 Epidemiology

We are unaware of convincing data which demonstrate as yet a likely connection between the above-described misuse of mill tailings and ill health among the exposed population. While a clear epidemiological connection exists between occupational exposure to radon daughters in air and risk of lung cancer and between exposure to other ionizing radiations and the risk of lung cancer (see Table 2.4, preceding chapter), there seem not to be significant recent increases in mortality or in the incidence of tumors among the population exposed to tailings in Grand Junction and elsewhere in the Southwest. The matter is under intensive study and includes other possible manifestations of irradiation, such as congenital defects, leukemia, and other non-respiratory cancers.[15]

Such a statistical survey is difficult because of the following factors: (a) A latency of 10 years or more is to be expected in radiation-induced lung cancer. (b) The concentrations of radon daughters differ greatly from time to time within a structure and widely among structures. The data required to estimate cumulative exposures are only now being collected. (c) The population is mobile in the fashion of Americans so that the exposed population changes rapidly. (d) The exposed population is very heterogeneous with respect to age, race, and other social factors. (d) The exposed population is relatively small (perhaps 5000) and the average accumulated exposure to lungs is small (perhaps 10-20 rad). The risk of lung cancer from radon daughter inhalation is about 1 per year per rad per million persons. Thus one might expect to observe less than one extra case of lung cancer per year among the exposed population. The normal incidence of lung cancer is about 2 per year per 10,000 persons.

3.6 Remedial Action

The remedial actions indicated are, of course, the removal of the tailings, improving ventilation, or erecting vapor-type barriers against radon. The object should be to bring most or all inhabited structures below an integrated level of 0.01 WL. One estimate for removal of tailings is in the range of $8 million for Grand Junction alone if the 0.01 WL criterion for action is used.[19] An upper bound would be something like $5000 over each of 7000 buildings for a maximum of $35 million. At this time little has been done to remedy the situation, but some $5 million in federal financial assistance is becoming available to limit radiation exposure from tailings around buildings.

The bulk of the one hundred million tons of tailings still remain in central piles; most are unstabilized or only partially stabilized against erosion. Several measures can be taken to minimize the radium hazard from mill tailings. The simplest is to flatten the piles as necessary and to cover them with earth. This cost about $0.3 million in the case of one demonstration pile of tailings,[10,13] and was effective in reducing the escape of radon and in absorbing γ-rays. An objection is based on the fact that the half-life of radium-226 is 1630 years. Thus, the radium is likely to outlive the protection of a thin earth covering. Earth covering applied to the 30-odd piles now in existence might cost $9 million. A more rational approach would be to return the tailings to the mines, since the radium content of the tailings would be nearly the same as that of the original uranium ore. The tailings could be mixed with cement and cast in place to ensure the entrapment of radon. A similar method could be used to link the disposal of uranium tailings to the restoration of land following the strip or open pit mining of almost any material.

A sophisticated and truely final solution to the tailings problem would be to chemically recover the radium (which has a chemistry similar to strontium and barium), incorporate it into reactor fuel, and finally destroy it by transmutation and fission. This is our recommendation regarding waste actinides in Chapter 8. The mass of radium involved is actually rather small. Assuming that existing tailings contain 700 pCi radium/g tailings (1 g radium = 1 Ci), corresponding to an initial uranium content of about 0.25%, then there would be 70 kg of radium in the one hundred million tons of tailings.

The burnup of radium is feasible in a thermal neutron reactor, such as the present light water uranium reactors used in power production. The first neutron absorption step

$$\underset{1630\ \text{yr}}{\text{Ra-226}} + \text{neutron} \rightarrow \underset{41.2\ \text{min}}{\text{Ra-227}}$$

has a thermal neutron cross section of about 20 barns compared with a cross section of about 3 barns for U-238, 6 barns for U-236, 30 barns for Th-230, and 50-500 barns for many isotopes of the higher actinides.[20] Through β-decay

$$\underset{41.2\ \text{min}}{\text{Ra-227}} \xrightarrow{\beta^-} \underset{22\ \text{yr}}{\text{Ac-227}}$$

radium enters the actinide series, where thermal neutron absorption becomes generally easy and where the fissionable isotopes Th-229, U-232, and U-233 are eventually built up.[21]

Extraction and purification of radium may be expected to double the cost of the chemical extraction and purification of uranium from ore and to add up to about 10% to the cost of uranium oxide. Radium burnup could be added to an established actinide burnup program at negligible additional cost.

It is not entirely clear what the AEC and/or state agencies will do with the vast quantities of tailings soon to be produced under the nuclear power program. According to one source,[4] all liquid, solid, and radiological wastes (including tailings) are to be discharged into tailings retention ponds at the mill sites. Eventually the ponds will be "graded, covered with earth, and restored for limited use." Grazing probably would be restricted. <u>We view the existence of such ponds, where tailings would contact acidic water for some 20 years and where groundwater pollution is possible, as totally unsatisfactory</u>.

3.7 Administrative Responsibility

The pollution of streams and sediments with radium from milling operations was clearly the result of the failure of the AEC to regulate a licensed operation. The production of radium-containing liquids and slimes and their dilution in local streams was considered a normal aspect of uranium milling. The AEC failed to anticipate important secondary effects - the sequestration of radium in flora and fauna; the rapid leaching of radium from fine mineral particles in water; the transfer of mill tailings into streams through erosion and flooding.

The problem of the mill tailings, used for fill and construction, represents, however, another example of purposeful evasion of responsibility by the AEC. Just as with the uranium miners, the AEC has defined its authority itself in such a way as to escape responsibility.[15] The production of tailings is considered a normal aspect of uranium milling, and so the piling of tailings on the mill sites and supervision of the piles were regulated by the AEC. How then could tailings be removed from the mill sites? Three loopholes existed. First, the AEC decided in 1966 that the regulation of tailings at the mill would terminate with the mill license - i.e., the tailings would be abandoned. Colorado promptly took control of unsupervised piles, but many remained abandoned elsewhere. Second, while the accumulation of radioactive tailings at mill sites was fully regulated, the transfer of tailings off the site for other uses was not viewed (at least after 1961) as an integral part of the operation and was not <u>subject to AEC regulation</u>, i.e., removal was not proscribed and the AEC never issued noncompliance citations to mill operators. Third, the AEC in 1947 defined a "source material" as an ore having more than 0.05% by weight of uranium plus thorium. Thus tailings, <u>once removed from a mill site</u>, were released from regulation by definition. In addition, the AEC claimed no authority over radium and radon off the mill site since these are not man-made radioactive elements. While they were not made by man, their radiological hazard was caused by man as a direct result of an AEC operation.

Taken together, this set of legalistic constructions relative to tailings cannot inspire confidence and represents little more than another transparent attempt to rationalize bureaucratic failure.

3.8 Occupational Hazard to Uranium Mill Operators

There is a growing body of evidence to suggest that exposure to uranium ore dust is more hazardous than previously suspected. A continuing study initiated in 1950 on 662 mill workers now suggests an excess mortality from lymphatic cancers.[22] Overall mortality was normal through 1967 (104 deaths vs. 105 expected), but 4 deaths due to lymphatic cancers were recorded vs. 1 expected.

Animal studies demonstrate that tracheobronchial lymph nodes tend to be the site of greatest concentration for inhaled uranium and thorium, and that thorium is very strongly retained by this tissue.[23,24] Thus, the lymphatic dose from thorium-230 would be about 20 times greater than that from uranium-238 and uranium-234 together. (These 3 isotopes would be present in about equal curie amounts due to secular decay equilibrium in uranium ore.)

Using an indirect and very approximate method, Archer et al.[22] estimate that the study group was occupationally exposed to Thorium-230 concentrations of about 4×10^{-11} $\mu Ci/cm^3$ of air for periods of 5-13 years in the case of the 4 men who died of lymphatic cancers. This can be compared with the maximum permissible concentration for Thorium-230 of 4×10^{-12} $\mu Ci/cm^3$ of air recommended by the International Commission on Radiological Protection (ICRP). If the excess mortality from lymphatic cancers is <u>in fact</u> due largely to inhaled thorium-230, then there would appear to be little safety factor in the ICRP standard.

Appendix 3.1

The letter allegedly sent from the AEC to the Departments of Public Health of Colorado, Utah, Wyoming, New Mexico, Arizona, Oregon, Washington, South Dakota, and Texas. The letter acknowledges a possible hazard in the misuse of uranium mill tailings. Source: ref. cited in note 15, pp. 24-25.

(Copy of letter sent to State Health official for each State.)

U.S. Atomic Energy Commission,
Washington, D.C., March 7, 1961.

Enclosed is a copy of the Commission's revised regulation, 10 CFR Part 40, "Licensing of Source Material", that was published in the Federal Register on January 14, 1961, to become effective on February 13, 1961.

Recently questions have arisen as to the applicability of AEC regulations to the control of sand tailings from uranium mill operations. These tailings contain small quantities of Radium 226 which must be considered in evaluating any health and safety problems which may arise in the uncontrolled use of the sand tailings.

The regulations define source material as follows: "Source material means (1) uranium or thorium, or any combination thereof, in any physical or chemical form or (2) ores which contain by weight one-twentieth of one per cent (0.05%) or more of (i) uranium, (ii) thorium or (iii) any combination thereof. Source material does not include special nuclear material."

The regulation also provides in Section 40.13 (a) that "Any person is exempt from the regulations in this part and from the requirements for a license set forth in Section 62 of the Act to the extent that such person receives, possesses, uses, transfers, delivers, or imports into or exports from the United States source material in any chemical mixture, compound, solution, or alloy in which the source material is by weight less than one-twentieth of 1 percent (0.05%) of the mixture, compound, solution or alloy."

Under the foregoing provisions, the milling of ores containing 0.05% or more by weight of uranium or thorium, or any combination thereof, must be licensed by the Commission and the milling operations must be carried out in accordance with Commission regulations. Since the discharge of radioactive mill effluent into the environment and the accumulation of radioactive tailings at mill sites are an integral part of the milling operations, such discharge and accumulation are, as you know, within the Commission's regulatory jurisdiction and are regulated by the Commission, notwithstanding the fact that the effluents and tailings generally do not contain 0.05% or more of uranium or thorium.

The transfer of sand mill tailings for uses such as road building, would not, however, be within the Commission's regulatory control since such uses would not be an integral part of mill operation. We are calling this fact to your attention because, as noted above, the radium content of these tailings may be such as to warrant control by appropriate state authorities over particular uses of tailings having a significant radium content.

We shall be glad to furnish any information or appropriate assistance to you in any activities which you may undertake to control proposed uses of sand tailings from uranium milling operation.

Sincerely yours,

H. L. Price,
Director, Division of Licensing and Regulation.

Enclosure: 10 CFR 40.

[Reprinted from 26 F.R., 284, January 14, 1961]

Notes and References

1 The γ-ray levels were measured and reported in terms of roentgen (R) units, which is the amount of radiation leading to the absorption of 88 ergs of energy per g of air. The unit which applies to tissues is the rad, corresponding to the absorption of 100 ergs per g of tissue. It has been determined that the exposure of tissue to 1 R of X- or γ-rays leads to the absorption of about 100 ergs per g tissue. Thus, the R and rad are numerically equivalent for X- and γ-rays. We shall use the rad unit throughout. Abbreviations used in this chapter include mrad (millirad) for one-thousandth (10^{-3}) of a rad, and μrad (microrad) for one-millionth (10^{-6}) of a rad.

2 BEIR Report: The Effects on Populations of Exposure to Low Levels of Ionizing Radiation, Division of Medical Sciences, NAS-NRC, Washington, D.C., 1972.

3 Working Level (WL) is a measure of the concentration of radon daughters in air. Working Level Month (WLM) is an index of occupational exposure (170 hours per month) to radon daughters. One WLM is equal to about 2 rads or about 6 rems to the lungs. See Chapter 2 and Table 3.1 for further discussion of units.

4 Environmental Survey of the Nuclear Fuel Cycle, Fuels and Materials Directorate of Licensing, USAEC, Washington, D.C., 1972.

5 See the preceding chapter for a description of the exposure of U.S. uranium miners to radiation.

6 The abbreviation pCi means "picocurie" or one trillionth (10^{-12}) of a curie. See Table 3.2 for the decay scheme for U-238.

7 Radioactive Water Pollution in the Colorado River Basin, Subcommittee on Air and Water Pollution, Senate Public Works Committee, Hearings, U.S. Congress, Washington, D.C., May 1966.

8 Uranium Wastes and Colorado's Environment, Colorado Department of Health, Denver, 1971.

9 The limit is 3 pCi of Ra-226 per liter.

10 H.P. Metzger, The Atomic Establishment, Simon and Schuster, New York, 1972, p. 163.

11 W.B. Harris, A.J. Breslin, H. Glauberman, and M.S. Weinstein, Environmental Hazards Associated with the Milling of Uranium Ore, HASL-40, USAEC, Washington, D.C., 1958.

12 Winchester Laboratory Reports WIN-112, -124, -125, by Natl. Lead Co. for USAEC, 1960 and 1961.

13 Evaluation of Radon-222 Near Uranium Tailings Piles, USDHEW, USPHS, Bur. Radiolog. Health, Rockville, Md., 1969.

14 Gamma rays arise in mill tailings from the disintegration of R-226, of the short-lived radon daughters Pb-214 and Bi-214, and of Pb-210. The short-lived radon daughters of Table 3.2 are Po-218, Pb-214, Bi-214, and Po-214. R-222 and radon daughters appear in the air because radon is a chemically inert

gas.

15 Use of Uranium Mill Tailings for Construction Purposes, Subcommittee on Raw Matls., Joint Committee on Atomic Energy, Hearings, U.S. Congress, Washington, D.C., October 1971.

16 Telephone interviews of June 27, 1973, with Lee Grossman, Colorado Dept. of Health, Denver, and with David Duncan, project director of the Uranium Mill Tailings Study, Las Vegas Laboratory, EPA; Comprehensive Indoor Radon - Working Level Report (Colorado), EPA, 1973; individual state reports of gamma radiation surveys (Arizona, Colorado, Idaho, New Mexico, Oregon, South Dakota, Texas, Utah, Washington, Wyoming), EPA, 1973; computer data printouts of these surveys, available through EPA, 1973.

17 This study was carried out by the Las Vegas Laboratory of EPA for the Colorado Department of Health.

18 See pp. 80-82 of ref. cited in note 15.

19 Removal of Uranium Mill Tailings from under or near Certain Residences in the Grand Junction, Colorado, Area. Report to the USAEC from Nelson, Haley, Patterson, and Quirk, Eng. Consultants, Denver, 1971.

20 O.R. Frisch, ed., The Nuclear Handbook, Van Nostrand, New York, 1958.

21 H.C. Claiborne, Neutron Induced Transmutation of Elements, ORNL-TM-3964, Oak Ridge, Tenn., 1972.

22 V.E. Archer, J.K. Wagoner, and F.E. Lundin, J. Occupational Med. 15 11 (1973).

23 L.J. Leach, E.A. Maynard, H.C. Hodge, et al., Health Phys. 18 599 (1970).

24 B.O. Stuart and T.M. Beasley, Selective Tissue Accumulation of Uranium and Thorium in Rats After Inhalation of Uranium Ore Dust, BNWL-122. Hanford Biology Research Annual Report for 1964, pp. 21-24, USAEC, Richland, Wash., 1965.

25 Report prepared by R.D. Sick, J.B. Baird, R.N. Snelling, D.L. Duncan, and P.B. Smith, Uranium Mill Tailings Problems in Grand Junction, Colorado, Aug. 1971.

Chapter 4

CATASTROPHIC NUCLEAR REACTOR ACCIDENTS

Daniel F. Ford, Henry W. Kendall

4.1 Introduction

Very large quantities of radioactive materials are necessarily associated with the operation of nuclear power reactors. These extremely hazardous materials are the "ashes" from the nuclear "burning" of uranium fuel. The problem of nuclear reactor safety is the problem of assuring that these radioactive materials, inescapably associated with uranium fission, will not be allowed to escape inadvertently into the environment. The release of only a small fraction of a commercial reactor's inventory of radioactive materials could set the stage for a catastrophe of unprecedented scale. A major controversy has now developed over the credibility of the assurance that the chances of a major accident are sufficiently remote so that the risks presented by the expanding nuclear power program are acceptably small. Especially important are several unresolved questions on whether required emergency systems will be able to mitigate and control potentially serious accidents.

4.2 Principles of Reactor Operation

A nuclear reactor is a device to establish and control a fission chain reaction utilizing uranium or some other fissionable material. Heat produced by the fissioning (or splitting) of the uranium nuclei is used eventually to heat water. The resulting steam, just as in a coal- or oil-fired power station, drives steam turbines that in turn spin electric generators. The two principal types of reactors in use in the United States circulate water through the reactor as the means of carrying off the heat generated by the nuclear fission in the reactor's core. In the boiling water reactor (BWR), water within the core is allowed to boil, at pressures near 1000 pounds per square inch (psi), and to create steam within the reactor itself. In the pressurized water reactor (PWR), the water is prevented from boiling by operation at pressures near 2200 psi; this water carries the heat to a secondary water system where steam is generated. (See Figure 4.1) In water cooled reactors, whether of the boiling water or pressurized water type, water temperatures are around 600°F. Other types of reactors use different types of coolants, such as helium, carbon dioxide, or liquid sodium, to transfer heat to a secondary water system. This chapter is concerned with water cooled reactors, the principal commercial reactors in the United States, and excludes consideration of other designs.

4.3 The Worst Possible Nuclear Accident

Laymen might expect that a nuclear explosion could occur in some abnormal circumstance at a nuclear power plant, but this is not possible. Fissionable uranium represents only about 3% of the uranium in a power reactor and is too diluted to explode. The most serious accident a reactor can undergo is one that involves overheating and melting of the reactor core, accompanied by leakage of radioactive materials into the surrounding area. An accident that developed to melting would involve breaching of all man-made structures

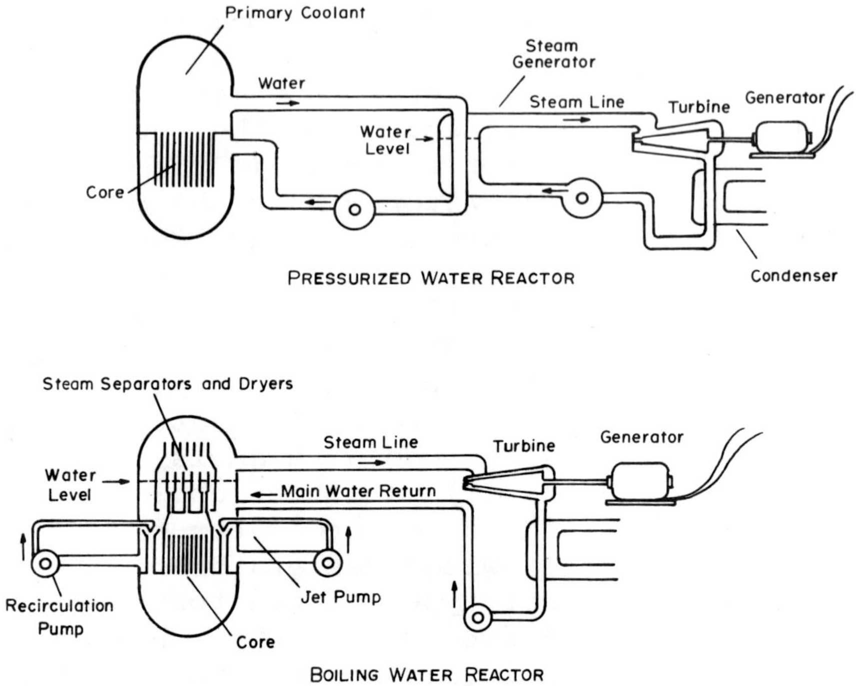

Figure 4.1 Schematic diagrams of a pressurized water reactor (PWR) and a boiling water reactor (BWR). In a PWR the water boils only in the secondary system. In a BWR water boils in the reactor vessel itself.

with more than 100 tons of white-hot radioactive material melting its way hundreds of feet into the ground. The molten mass would include billions of curies of radioactivity, more than one-fifth of it gaseous or volatile. A large reactor prior to a periodic refueling will contain about two tons of radioactive material mixed within about 100 tons of slightly enriched uranium dioxide. For comparison the Hiroshima atomic bomb produced about two pounds of similar material. The accident is colloquially referred to in the nuclear industry as the "China syndrome," reflecting the direction, although not the destination, of the molten material. Release of very large quantities of radioactivity appears highly likely in a "China" accident owing to the difficulty of retaining the gaseous materials as the core melts downward.

The starting point for the "China syndrome" is an abrupt loss of normal cooling water as would occur, for example, following a major pipe or reactor vessel rupture. These ruptures could result from defects in reactor component design, quality control deficiencies, welding errors, overpressurization during transient abnormal reactor operation, or from events such as earthquakes or sabotage.

At the onset of a loss-of-coolant accident the chain reaction automatically halts. This is because the cooling water that is normally present provides material necessary to slow down or "moderate" neutrons emitted in fission. The "moderated" neutrons can then initiate further fissions, thus sustaining a chain reaction. Cooling water loss, even without emergency power shutdown from control rod insertion, terminates the chain reaction. However, there is another source of heating in the reactor core that continues and cannot be turned off: heating provided by the accumulated radioactive wastes. This heat source alone can induce soaring fuel temperatures in the absence of adequate cooling and is more than sufficient to precipitate core meltdown. If means for removing this heat are not provided, a loss of coolant accident will develop out of control within minutes of the initiating rupture. Breach of the reactor structures will then surely result within a few hours.

Radioactive heating, often called "decay heating" because it arises from the energy release accompanying the decay of fission products, can develop over 200 megawatts of thermal power at the instant of rupture for a typical large nuclear power plant. The decay heat generation drops to about 60 megawatts after 15 minutes. Thereafter, the heating decreases relatively slowly for a period of weeks before starting a more rapid decline.

4.4 The Emergency Core-Cooling Systems (ECCS)

Loss-of-coolant accidents have always been considered credible events. The AEC accordingly requires all power reactors to have emergency core-cooling systems (or ECCS's, as they are commonly called) to provide timely and adequate cooling to arrest the course of accidents and to prevent core meltdown. Presently, controversies revolve about both the chances of a pipe rupture or other event causing reactor loss of coolant and whether or not the ECCS would prove effective in controlling the resulting accident.

Concern within the AEC over the effectiveness of ECCS safety systems began in 1965 as reactor sizes increased dramatically into the two to three thousand thermal megawatt range (see Figure 4.2).[2] Each such large reactor allows the generation of from 700 to 1000 megawatts of electricity. The remaining energy, 1400 to 2000 megawatts (thermal), is discarded as waste heat. Both the possible severity and the difficulty of gaining control of accidents were known to increase

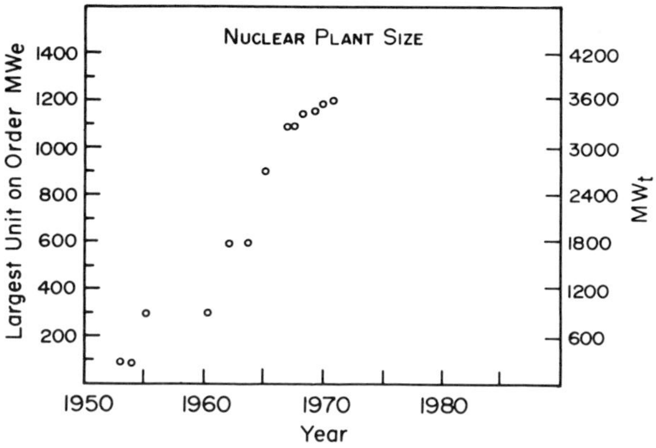

Figure 4.2 Growth in the size of new nuclear power plants between 1950 and 1974. Source: WASH-1250, p. I-6.

greatly for larger reactors.

The AEC appointed a task force in 1966 under the late W. K. Ergen at Oak Ridge National Laboratory, one of the AEC's important centers for reactor safety investigations, to study the ECCS problem. The AEC also requested that laboratory and Battelle Memorial Institute to investigate specific research needs relating to loss-of-coolant accident control. All of these early studies indicated that there was urgent need to achieve experimental confirmation of ECCS effectiveness.

As a consequence of the Ergen Task Force recommendations, Oak Ridge's Reactor Safety Division undertook studies of fuel element failure modes and certain important heat transfer phenomena. Experimental and analytical studies of ECCS cooling capability were also initiated at what has become the AEC's prime

ECCS research laboratory, the National Reactor Testing Station in Idaho operated by Aerojet Nuclear Company under AEC contract. Some studies employing full length electrically heated fuel bundles were carried out. The full length fuel bundle tests were actually done by the two major reactor manufacturers, Westinghouse and General Electric, under subcontracts monitored by AEC consultants at the Idaho laboratory.

The AEC also began to formulate a comprehensive safety program to resolve the concerns. This plan,[3] appearing in early 1970, represented a consensus of industry and AEC opinions regarding safety research priorities. It identified "all of the factors affecting ECCS effectiveness" as "the most urgent problem areas in the safety program today," and pinpointed a number of specific problems as "very urgent, key problem areas, the solution of which would clearly have impact, either directly or indirectly, on a major critical aspect of reactor safety."

In the fall of 1970 a series of tests was run at the National Reactor Testing Station using a nine-inch-diameter model reactor with electrically heated simulated fuel rods. The purpose of the series was to evaluate some of the important effects of loss of cooling water and subsequent injection of cold emergency coolant in a pressurized water reactor. The model was not a precise simulation of a large power reactor, although its configuration and expected behavior were believed well understood and thus predictable. The model emergency system unexpectedly failed in all of the tests to deliver water to the core, and the AEC's computer program that mathematically modeled loss-of-coolant accidents failed unexpectedly to predict critical test results with the necessary precision. The same computer models were being used to determine ECCS effectiveness in the licensing of commercial nuclear power plants.

Other aspects of the AEC's research program had been progressing far slower than had been hoped for and some critical programs had been terminated short of completion.[4] Resolution of an important number of the issues identified in the Ergen report continued to be delayed. Certain studies, identified as urgent by senior AEC personnel, were deferred. The Advisory Committee on Reactor Safeguards (ACRS), the AEC's watchdog committee, has continued to recommend "substantial" increases in funds for research into a large number of important aspects of accident phenomena, several of them still not resolved after many years.

4.5 The ECCS Interim Acceptance Criteria

These test failures and other concerns troubling the safety researchers led the AEC Director of Regulation in early 1971 to convene a task force to review ECCS behavior in light of the rapidly increasing number of large reactors coming into operation. Based on the recommendations of its task force, in June 1971 the AEC prepared revised safety rules, the Interim Acceptance Criteria, and "in view of the public health and safety considerations" found that the Criteria "should be promulgated without delay."[5]

The present authors with two of their colleagues in the Union of Concerned Scientists prepared analyses both of the implications of the unexpected test failures and of the Interim Acceptance Criteria.[6] The analyses resulted in the conclusion that the Criteria provided inadequate assurance of safety. An important consequence of these analyses was a major AEC hearing on ECCS[7] with the AEC Regulatory Staff, each of the major reactor manufacturers (General Electric, Westinghouse, Combustion Engineering, Babcock & Wilcox), several electric utilities, and a citizens' group, the Consolidated National Intervenors, as participants. The Union of Concerned Scientists prepared and presented the Intervenors' technical case. Although scheduled to last no more than six weeks, the hearings extended over some twenty months, from January 1972 to October 1973. A record of oral proceedings of 22,000 pages accumulated, along with an equal quantity of technical documents of record. It is in this hearing record that the safety controversy is most sharply documented - in written and oral testimony, cross-examination, and in AEC internal memoranda, reports, analyses, and minutes of meetings, released publicly for the first time.

A report[9] surfaced during the ECCS hearing prepared by the AEC's principal ECCS consultants at Aerojet Nuclear Company that treated in detail 28 areas of knowledge and analytical capability important to pressurized water reactor ECCS operation. Seven of the techniques for dealing with these areas were identified as completely missing. The status of the remainder were variously identified as "incomplete," unverified," "inadequate," "imprecise," "uncertain," or "preliminary." Eight areas were designated as of very high urgency, 15 of high urgency.

Among other documents was a long letter[20] to the AEC from the Director of the Oak Ridge Nuclear Safety Program setting forth the views of a number of that group's safety personnel. It included the statement:

"To summarize what follows herein, we are not certain that the Interim Criteria for ECCS adopted by the AEC will, as stated in the *Federal Register* 'provide reasonable assurance that such systems will be effective in the unlikely event of a loss-of-coolant accident.'"

Cross-examination identified over two dozen AEC safety researchers who had misgivings about the effectiveness of ECCS. These and other evidence contrasted[11] with the AEC's official position, strongly endorsed by the reactor manufacturers, that "the conservatism in today's criteria and today's (computer) evaluation models suffice to provide reasonable assurance of the safety of today's reactors."

4.6 The Consequences of a Nuclear Accident

The importance of ECCS effectiveness is measured by the size and scale of the possible consequences of reactor core meltdown.

Breaching of the reactor structures in the course of core meltdown could release to the environment an appreciable fraction of the reactor's radioactivity that is gaseous or easily volatile, and possibly other less volatile elements as well. The material, if carried off by wind under temperature inversion conditions that would cause meteorological trapping, would be lethal at distances up to

15 miles, with potential damage extending to hundreds of miles, according to a 1957 study carried out by the Brookhaven National Laboratory for the AEC.[12] For this hypothetical accident, WASH-740, as the Brookhaven report was designated, estimated 3400 people killed and 43,000 nonlethal injuries. This accident envisaged the most severe combinations of circumstances that the authors considered credible. More probable combinations led to less severe results.

The Brookhaven study's calculations were, however, based on an assumed reactor size of 500 Mw, thermal, a very small reactor compared to the 3600 Mw reactors now under construction in the United States. Moreover, the example reactor in the Brookhaven study was on a shorter fuel cycle than is now common, i.e., the radioactive wastes were removed more frequently.

The present authors' analyses for the most severe accident in a large, modern reactor indicated a maximum lethal range in excess of 65 miles and radioactive damage or fallout possible at distances beyond 400 miles under temperature inversion conditions.[13] A second and secret AEC update of WASH-740,[14] carried out by Brookhaven National Laboratory in 1964-1965, more than confirmed these estimates. This latter study stated that a major accident might create an "area of disaster...equal to that of the State of Pennsylvania." The updated study, not released until 1973, showed example calculations indicating 45,000 fatalities from an uncontrolled loss-of-coolant accident.

The most severe accident is regarded as less probable than an accident involving a smaller radioactive release or a less unfortuante combination of circumstances. Nevertheless, with reactors being sited within 10 or 20 miles of major urban areas, it is evident that a major accident could have consequences dwarfing any known peacetime event - an accident that the nuclear program itself would likely fail to survive. Great burden is thus placed on the effectiveness of emergency systems to control a loss-of-coolant accident, and an extreme diligence in the construction and operation of reactors is required to reduce the chances that loss-of-coolant accidents will occur.

4.7 The Details of Reactor Design

In what follows we will outline, as an illustration of the basic ECCS issues, the development of a loss-of-coolant accident in a pressurized water reactor and the phenomena that would act to make control difficult. Some of the problems are common to both boiling water and pressurized water reactors, others are not. The safety of both types is subject to controversy.

Pressurized water reactors may have two, three, or four "loops" in which reactor cooling water circulates, each with its own steam generator and circulation pump. Figure 4.3 shows the water flow in a two-loop plant; Figure 4.4, the arrangement of components in a typical four-loop plant. Each loop will circulate nearly 90,000 gallons per minute of primary coolant through the core when in full operation.

The nuclear chain reaction occurs in the uranium fuel that comprises the reactor core. The core is located in a large pressure vessel to contain the high pressure cooling water

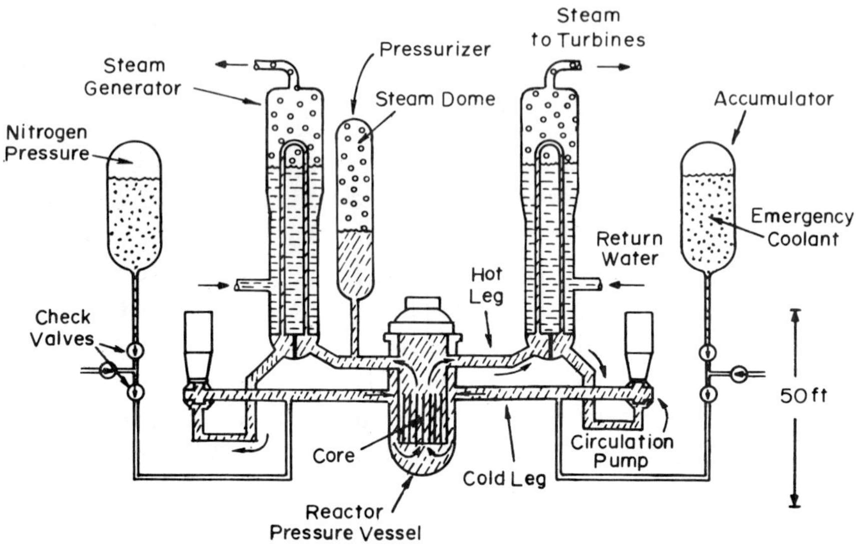

Figure 4.3 Normal water flow in the "primary" loops of a two-loop pressurized water reactor. The water is kept under great pressure in the primary loops and does not boil. Steam is produced in the steam generators and passes through the "secondary" loops to the turbines.

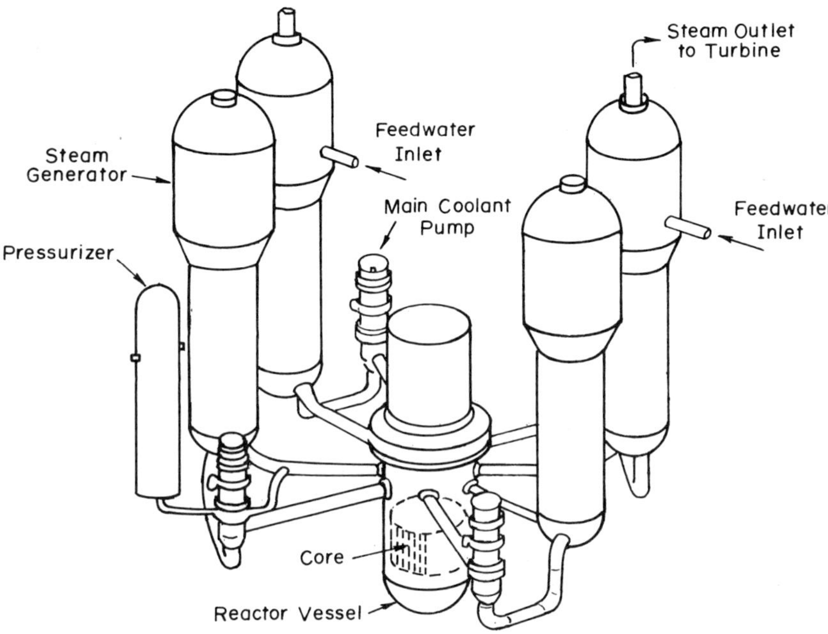

Figure 4.4 The arrangement of major components of the steam supply system of a four-loop pressurized water reactor.

(see Figure 4.5). The core, some 12 to 14 feet in diameter, is composed of 30 to 40 thousand fuel rods, each 12 feet long and about one-half inch in diameter. The rods are arranged in an open lattice configuration about one-half inch apart and are bathed in the circulating primary cooling water which carries off the generated heat. Channels are left in the core to allow for the insertion of neutron absorber rods to control the rate of the chain reaction so as to alter the reactor's power level. The fuel rods have an external cladding of an alloy of zirconium, chosen for its favorable nuclear properties. The purpose of the cladding is to seal in the uranium fuel to prevent release of radioactivity into the cooling water, and to establish a fixed geometry for the otherwise loose fuel pellets and thereby assure efficient heat transfer. The fuel pellets are small, very dense cylinders of sintered uranium dioxide, with several hundred stacked inside each rod. There are nearly 10 million pellets in a large reactor. Fuel rods generate different amounts of power, depending on their position in the core and their operating history. Although the average linear power generated is close to 7 kilowatts per foot, some rods can develop as much as 18 kilowatts per foot with internal temperatures in the uranium dioxide pellets approaching 4000°F. Outside surface temperatures of the fuel cladding are all near 650°F.

The reactor pressure vessel, primary-loop coolant pumps, and steam generators are all contained within a massive, reinforced concrete structure with a steel liner, called the containment (see Figure 4.6). The containment is itself a pressure vessel. It is designed to retain the entire primary coolant

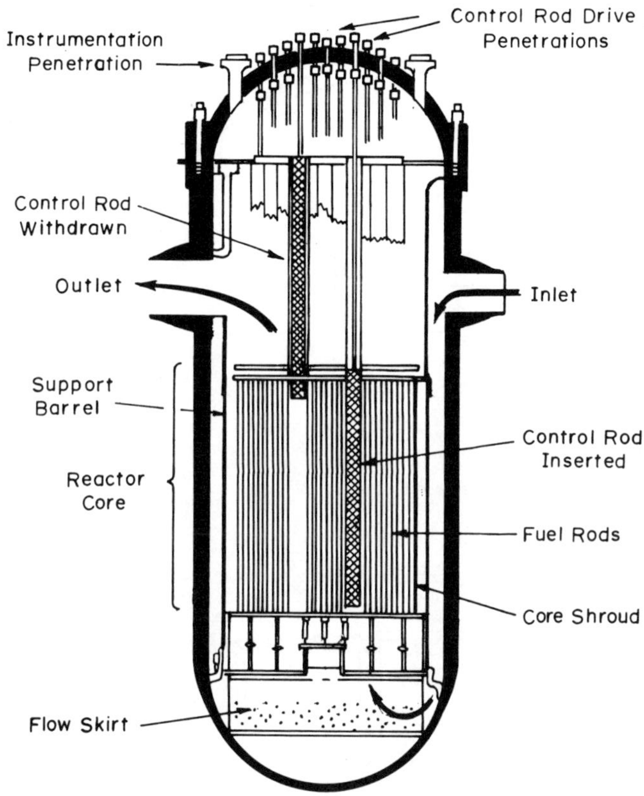

Figure 4.5 Cutaway view of the reactor vessel in a pressurized water reactor. The vessel is over 40 feet high, with an inside diameter of over 14 feet and wall thicknesses in excess of 8 inches. There are over 40,000 fuel rods in the vessel, containing more than 100 tons of uranium dioxide.

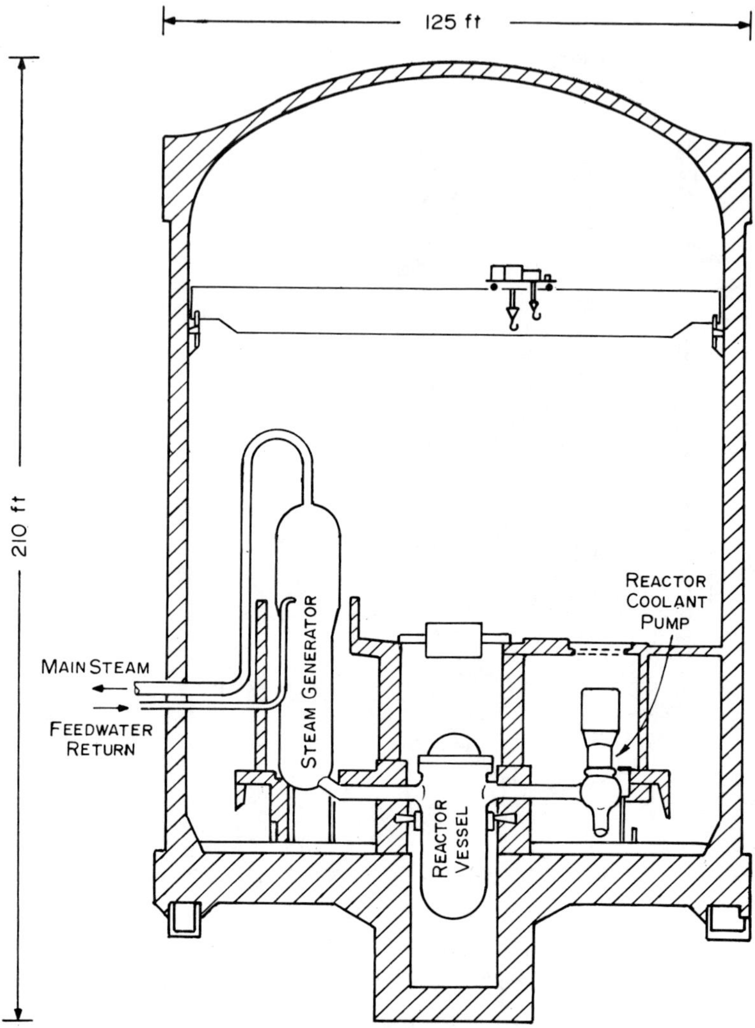

Figure 4.6 The reinforced concrete containment building with a steel vapor liner, for a pressurized water reactor.

in the event of a loss-of-coolant accident induced by a rupture of one of the primary loop recirculation pipes. The design is such that, if a loss-of-coolant accident is brought under prompt and continuing control by the ECCS, whatever radioactivity might be released from the fuel rods would not reach the environment in damaging amounts but, instead, would be trapped inside the containment to decay harmlessly away.

4.8 Loss-of-coolant Accidents

In the accident analysis required by AEC for each reactor, loss of coolant through a pipe rupture must be considered. Analyses considering various locations where such a break could occur indicate that a break in a return line, or "cold leg," of one of the 24-inch primary recirculation lines, would be the worst (see Figure 4.3). This cold-leg break is regarded by the AEC as the most severe accident that can occur in a nuclear reactor.

Reactor pressure vessel rupture could also lead to abrupt loss of coolant. The probability of such a rupture is inadequately known. In 1971 the AEC stated that the need for information on primary system integrity "both to assure safety and to enhance public confidence in the integrity of reactor systems is urgent." Professor David Rose, an MIT nuclear engineer, stated that though the reactor vessel "is made with the very best technology, more uncertainty exists about its quality, because of the nature of the thing. But also no protection could be envisaged for certain hypothetical failure modes of the reactor pressure vessel. Such failures have by edict been declared incredible." In 1965 the Advisory Committee on Reactor Safeguards, the AEC's "watchdog" committee, recommended that provisions be made against such failure and in addition that means be sought to ameliorate the consequences, recommendations the Committee repeated in 1973. However, no provision is made in modern reactors for coping with pressure vessel ruptures, nor are containments known to be strong enough to withstand the surging pressure and missiles developed in such an event. Recently an AEC official stated[15] that it was the inability of designers to construct an adequately strong containment that led to the decision to accept the risk of vessel rupture.

In Britain, a proposal was made by the Central Electricity Generating Board in 1973, supported by Britain's nuclear reactor manufacturers, to purchase American nuclear reactors. The British Parliament's Select Committee on Science and Technology[16] opposed the proposal on the grounds that American reactors were unsafe. Britain's chief nuclear inspector, in testimony before that Committee, rejected U.S. claims about the safety of these reactors. The British government's Chief Scientific Advisor, the metallurgist Sir Alan Cottrell,[17] opposed the plans to construct U.S. reactors in Britain on the grounds that his studies had shown the reactor pressure vessels to be susceptible of catastrophic rupture without warning. In the face of all this evidence the British government adopted the policy that it would not buy U.S. reactors.[18]

Other accident mechanisms that could lead to loss of coolant involve malfunction of certain valves with concurrent failure of reactor shutdown mechanisms, failure of certain critical steam generator components, operator error, overpressurization during system transients, or design errors aggravating normally innocuous accidents. These events generally are not believed to have the potential severity of a cold-leg break or some vessel ruptures, but could still result in substantial radioactive releases if emergency systems failed. Not all of the spectrum of accidents has been fully analyzed.

4.9 A Major Pipe Rupture

What happens in the event of a cold-leg rupture in an operating pressurized water reactor? No one knows for sure because large scale well-simulated experiments have not been carried out. The following rough picture is based on present understanding. In the first half-second after the break, the vessel pressure is expected to drop very sharply as the elastic compression of the water is relieved. The flow of cooling water through the core itself would decrease rapidly as the superheated water flashed into steam and new flow patterns were established toward the rupture. Virtually all of the primary coolant would be expelled violently into the containment in perhaps as short a period as 10 to 15 seconds. During this rapid depressurization, or "blowdown," fluid flow through the core decreases and may oscillate and possibly halt briefly one or more times. The vessel would continue to depressurize as the last of the reactor water is vaporized and vented. Heat-up of the fuel rods starts as a result of the radioactive heating, redistribution of the high levels of thermal energy stored initially within the uranium, the decrease in flow of coolant, and poor cooling as a result of steam replacing liquid coolant. Increases in fuel cladding temperature would be in the range of 50°F to 100°F per second. Computer model predictions suggest that rod temperatures may rise above 2000°F, decrease briefly, and then climb again in ways difficult to understand thoroughly.

Emergency coolant is available in ECCS reservoirs, called "accumulators," to resupply the core with water. These reservoirs are passive devices, prepressurized with nitrogen, that can come into action automatically (see Figure 4.3). Each reactor cooling loop has one such reservoir which can deliver about 7000 gallons of water starting as the reactor pressure falls below 600 psi (see Figure 4.7), a pressure that is reached some 15 seconds or so after rupture. By this time in the accident the reactor water level has dropped below the core bottom and the vessel is expected to be nearly empty. Emergency water that is injected during blowdown into the intact loops as well as the ruptured loop will very likely be swept along with the primary coolant and lost out the break. It is only after depressurization is nearly complete that emergency water can begin to refill the vessel, a time not reached until about 25 seconds after rupture. There is presently a question whether thermal shock to the 600°F pressure vessel from the 120°F emergency coolant would induce fracture of the vessel. Some types of fracture with attendant new leakage paths - such as the bottom of the vessel - could make attempts to control the accident far more difficult.

For 10 seconds or more after the start of

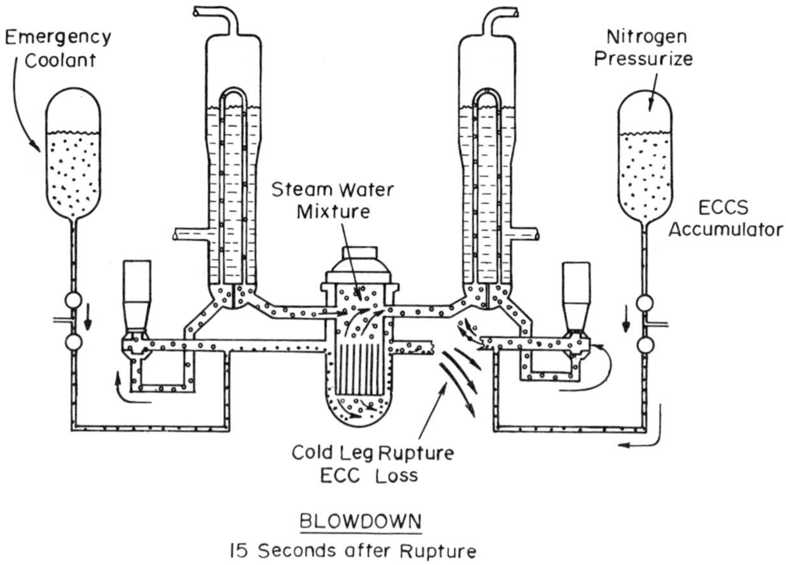

Figure 4.7 A two-loop pressurized water reactor, 15 seconds after a loss-of-coolant accident. Emergency core-cooling water begins to enter the reactor vessel from the accumulator in the unbroken side as soon as the system pressure falls below 600 psi.

emergency coolant injection, the core receives only feeble cooling from steam carrying some entrained water droplets. Temperatures in the hottest part of the reactor are still increasing at perhaps 30°F per second. No substantial cooling of the core can occur until the level of the emergency water reaches the bottom of the core, an event which is not expected to occur until some 20 to 25 seconds after the rupture, when some fuel rods may have reached temperatures in the vicinity of 2200°F to 2300°F. At this time steam and entrained water would - it is hoped - start to flow upwards in increased quantity through the core and then through the steam generator and pump of the broken loop to exit at one end of the ruptured pipe (see Figure 4.8). If the ECCS is able to reflood the hot core as it is designed to do, the water level will continue to move upward and terminate the fuel-rod temperature increase before the highest temperature in the reactor exceeds 2200°F, approximately 30 seconds after the rupture.

At about this time water in the reservoirs, or accumulators, is exhausted. Additional and continuous reflooding is expected to be then provided by a number of pumps driven by

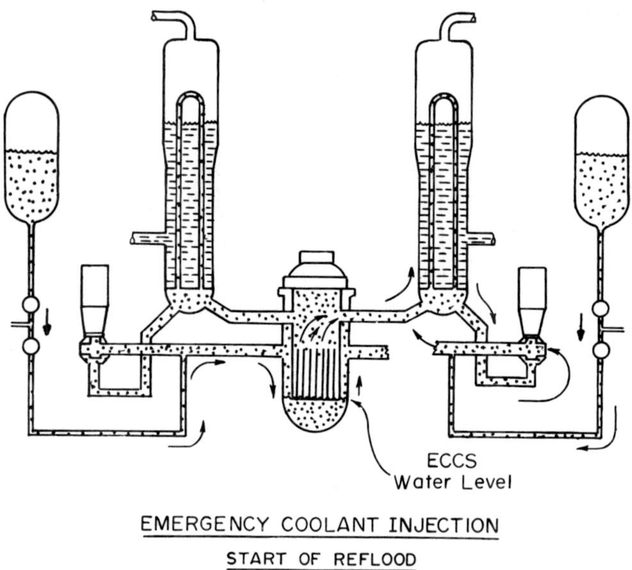

Figure 4.8 A two-loop pressurized water reactor, about 20 seconds after a loss-of-coolant accident. It is hoped that emergency coolant will begin to fill the reactor vessel and cool the core. The steam from the core must exit through the steam generator and pump in the broken loop.

normal or emergency power and drawing water from an array of external sources. _If_ the coolant wholly refloods the core, the entire transient is effectively over and full control of the accident has been gained. The risk of catastrophic accident is past provided only that continued circulation of cooling water through the core persists. As we shall see next, there are important reasons to question whether accident control may be effected in the way ECCS designers planned.

4.10 Consequences of an ECCS Failure

It is in the critical time period from about 15 seconds after the rupture to 30 seconds that control of the accident _must_ be gained by ECCS operation and the fuel temperature excursions halted. The greatest importance is attached to the need for having an adequate cooling water flow upward through the core in this period. If the vertical flooding rate, once emergency coolant reaches the core bottom, is below some critical value, presently believed to be in the vicinity of 0.7 inches per second, then the accident will proceed out of control.

4.11 Metal-Water Reactions

Loss of control can result from a new source of heat that develops if temperatures climb much above the 2000°F level for very long. This heat is provided by chemical reaction of the zirconium fuel-rod cladding with steam or liquid water. The reaction rate increases strongly with temperature, liberating hydrogen through oxidation of the alloy cladding of the fuel rods. In effect, the zirconium alloy cladding starts burning in the oxygen it frees from the water. Oxidation of the clad, in addition to generating heat, can cause severe embrittlement of the fuel-rod cladding and thus can enhance core damage. The loss of core geometry from fracturing can result in uncoolable configurations. The cladding-water reactions provide a significant addition to the radioactive heating at 2300°F, and become autocatalytic, that is, self-sustaining, at about 2700°F. Somewhere above this temperature, the continued application of emergency cooling water would aggravate rather than mitigate an accident because it would be providing oxygen for additional metal-water reactions and additional heat. Once this occurs, there is great risk that control over the accident will be irretrievably lost, resulting in a meltdown no longer capable of being arrested.

Temperatures in an uncontrolled accident would continue to climb. According to an AEC study, at 3300°F the fuel-rod cladding would start to melt with a further sudden increase in the metal-water reaction rate and the possibility of molten-metal-water explosions. Once uncontrolled temperature excursions and melting have started somewhere in the core, they can be expected to propagate and the damage will spread. Irrespective of the occurrence of explosions, hydrogen generation would continue and would put the integrity of the containment at risk.[19]

4.12 Core Meltdown

Core collapse would start above 3300°F and within ½ to 1½ hours, depending on the accident, the accumulating molten mass of fuel and core structure, at a temperature approaching 5000°F, would have melted through the pressure vessel, and, within a few additonal hours, through the concrete structure of the containment below the reactor. Material below man-made structures would not be expected to retain the hot radioactive materials adequately, especially the gases, and an unacceptable portion would be virtually certain to escape. Furthermore, additional metal-water reactions between the molten fuel mass and ground water below the nuclear plant would blow steam channels upward around the concrete base of the reactor containment building. These channels would become geysers venting radioactive materials out into the environment to become airborne. If the released materials approached only a few percent of a large reactor's routine radioactive inventory, as might well occur, a substantial catastrophe would be possible.

4.13 Computer Models of ECCS Performance

In order to show that a loss-of-coolant accident will be controlled by the ECCS, resort is made by the AEC and by the reactor manufacturers to computer calculations. There is a continuing lack of large scale experiments on which to base a thorough understanding of accident mechanisms. These computer programs or codes, as they are called, are indended to predict the course of the accident and, among other quantities, develop the expected temperature history of the fuel cladding. They amount to attempts to simulate by mathematical means the complex course of an accident in a large and very intricate machine.

The knowledge necessary to construct the codes and establish the values of the hundreds of parameters they require can in principle be gained in small-scale experimental studies of the processes relevant to the accident: determination of heat transfer to water, steam, and steam-water mixtures during blowdown; pressure-drop and fluid-flow experiments; reflooding heat-transfer determinations; pump responses to the disruptive and violent flows during blowdown; entrainment of water by steam violently produced at red- and white-hot surfaces; thermal radiation; the rate of decline of radioactive decay heating; nature of the heat transfer from the uranium fuel to the fuel cladding; and a host of others.

The thermal-hydraulic behavior of the coolant and the behavior of the core are extremely complex in accident circumstances. The validity of the code predictions of these complex phenomena is at the center of a major controversy. Very important criticisms relate to the quality and scope of the underlying experimental information: whether all the phenomena have been adequately replicated in the laboratory; and, even more important, whether they have all been identified and their effects correctly assessed. Of considerable importance, in addition, is whether the mathematical and conceptual approximations that have to be made in the course of formulating the codes are adequate to the task.

4.14 Weaknesses in the ECCS Codes

An AEC task force was organized in 1970 when the ECCS computer codes failed to accurately predict the accident behavior of a 9-inch semiscale model reactor. The task force commissioned the Aerojet Nuclear Company to prepare a detailed review of the state of the art of the codes and the validity of predictions they made of the complex phenomena encountered in a major accident. The two-volume report[20] that Aerojet Nuclear prepared emphasized how difficult it was to gain a clear understanding of the course of an accident. It concluded that, in several critical areas, the ability to predict what would happen was simply "beyond the present capability of engineering science."

Two senior AEC engineers who had worked on ECCS problems wrote to the task force in 1971[21] that "...the consummate message in the accumulated code outputs is that the system performance cannot be designed with sufficient assurance to provide a clear basis for licensing [reactors]." A later report prepared by the AEC[22] itself stated: "As reactor designs and their operating characteristics changed, the analysis methods were 'patched up,' rather than redeveloped with the net result that overall existing methods are inefficient, inflexible, and do not adequately represent the physical phenomena intended." An author of one of the very large codes relied on for safety analyses stated that an adequate code would need to be ten times bigger than any now running in order to accomodate the complexities involved, and spoke of present codes as inadequate, lacking verification, and unrealistic. Dr. Alvin Weinberg, the Director of Oak Ridge National Laboratory, wrote the Chairman of the AEC in 1972,[23] with respect to the use of the codes in reactor safety analyses, that he had a "basic distrust of very elaborate calculations of complex situations especially where the calculations have not been checked out by full-scale experiments." He emphasized the importance of carrying out a program of such measurements.

4.15 Phenomena Omitted from ECCS Codes

Many of the weaknesses in the codes on which stated ECCS performance rely have been identified following unexpected discovery or belated recognition, late in the reactor program, of phenomena that have been found to aggravate the task of controlling an accident.[24] Some of these phenomena are left unaccounted for in present computer codes.

The first of these omissions is called "steam binding," and arises when the level of the emergency cooling water reaches the bottom of the core. At this point some of the water vaporizes to form great quantities of steam with entrained water. This steam can leave the pressure vessel only through the ruptured loop and must pass through the steam generator and recirculation pump before venting through the break. The steam generator and pump both offer resistance to the steam flow (see Figure 4.9). If there were no inhibition to the reflooding, the water level would rise in the core at between 8 and 10 inches per second. In 1967, when the pressurized water reactor ECCS was designed, it was believed that rates close to these values would be achieved in practice. AEC consultants at Aerojet Nuclear, however, established that resistance offered to the steam flow was substantial and would act to reduce the predicted flooding rates: back pressure of steam above the core would inhibit reflooding of the core. As the vessel floods, the water level would rise in the inlet supply annulus around the core, although not within the core itself,

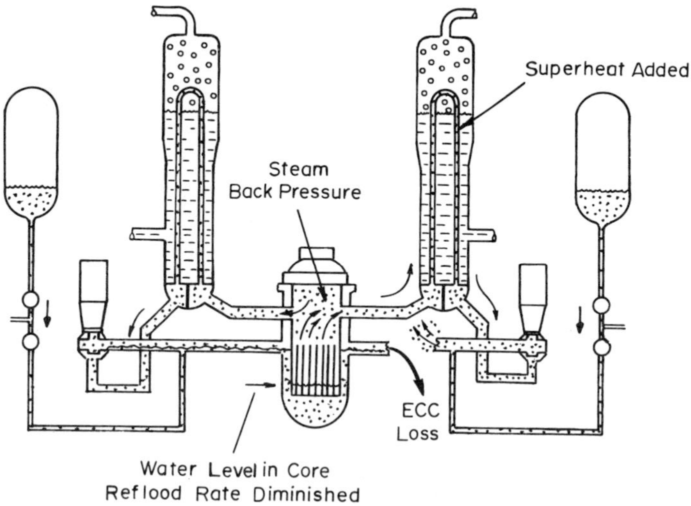

Figure 4.9 A two-loop pressurized water reactor, 25-30 seconds after a loss-of-coolant accident. The steam rising through the core must vent from the pressure vessel by passing through the steam generator and recirculation pump. The resistance offered to the steam is called "steam binding" and significantly reduces the reflooding rate.

until the coolant would spill out the rupture and be lost, as shown in Figure 4.9. Successive improvements in the analysis showed successive decreases in the predicted flooding rates.

Even further decreases in estimated reflooding rates occurred following recognition that the passage of steam and entrained water through the steam generator with its reservoir of heat would vaporize the entrained water and superheat the steam, causing further flow resistance and back pressure. At the present time, reflooding rates expected for an accident in an operating reactor are between 0.9 and 1.5 inches per second, values earlier considered as extremely degraded. Since an accident is expected to be uncontrollable when the rate is in the vicinity of 0.7 inch per second, there are, at best, relatively small safety margins between effective cooling and loss of control. One senior Oak Ridge reactor safety engineer testified at the ECCS hearing that a clear demonstration of coolability would be necessary to satisfy his uncertainties regarding ECCS capability. He said that cooling by narrow margins would have to be regarded as an essentially uncoolable situation. Other effects discussed below appear for some accidents to have eroded these narrow margins entirely to zero.

The steam binding phenomenon highlights the most critical design weakness in U.S. pressurized water reactors - a weakness that has made the margins of safety extremely vulnerable to new and unexpected accident events. This weakness is the absence of positive means of <u>forcing</u> large quantities of emergency coolant into the core, providing overwhelming and positive control of the radioactive heating, and thus ensuring that an accident can never approach uncontrollability. The maximum pressure available to move coolant upwards into the core arises from the difference in height, the pressure head, between the water level at the bottom of the core and the level of the broken pipe where liquid is lost. At most, the pressure available is about 7 psi and proportionately less as the water level in the core rises. Steam binding develops back-pressures that sharply oppose this feeble pressure. It is with good reason that the AEC's Advisory Committee on Reactor Safeguards has strongly, and repeatedly, recommended the development of improved ECCS's.[25]

There are additional difficulties. Two of them affect the core reflooding rate directly - steam-generator tube failure and flow blockage.[26] An additional source of steam can develop to aggravate severely the steam binding problem. If the sudden shock of the initial pipe rupture were to rupture some of the tens of thousands of tubes in the steam generators, superheated water under high pressure would be injected into the primary system at virtually the worst possible place, namely, into the pipes leading to the top of the pressure vessel (see Figure 4.10). Something like a dozen or so ruptured tubes in the steam generators would be expected to <u>entirely stall</u> the core reflooding. No assurance exists that the main pipe break itself will not induce such ruptures. Since corrosion-thinning and leakage of steam generator tubes is already plaguing a number of operating reactors - some reactors have already been shut down temporarily on this account - it is apparent that the contingent rupture of a fraction of one percent of the steam generator tubes in the brusque circumstances of a reactor accident could well occur.

The second difficulty also involves accident-induced damage: the development of obstructions in the reactor core itself.[27] As rod temperatures in an accident rise above 1200°F the fuel rod cladding progressively loses strength and internal pressure from helium, normally present to improve heat transfer, and accumulated fission gases, causes

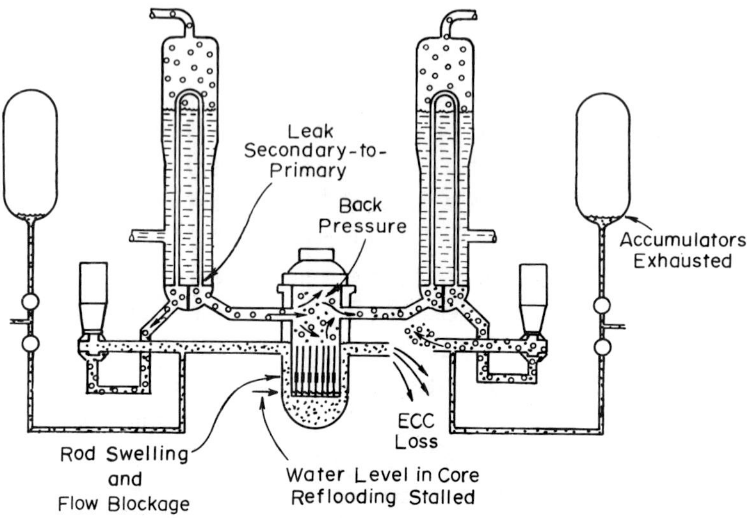

Figure 4.10 A two-loop pressurized water reactor, about one-half minute after a loss-of-coolant accident. If a dozen or more steam generator tubes rupture, superheated water will be injected from the secondary loop into the pressure vessel, likely preventing reflooding. Rod swelling can also exacerbate the accident by blocking the cooling water from the hottest part of the core.

fuel rods to swell and rupture. This occurs within ten seconds of pipe rupture in a pressurized water reactor and somewhat later in a boiling water type. A significant fraction of the reactor's fuel rods may be involved. A major program had been started at Oak Ridge National Laboratory following the Ergen report to study fuel rod swelling and attendant blockage, but was terminated short of completion in 1971. The results of the partial program indicated that under some circumstances the potential existed for nearly complete flow blockage to occur at least over some portions of a reactor core. The consequence could be to alter significantly the course of an accident. Figure 4.11 shows a cross section of a simulated bundle of fuel rods subjected to testing in this program. The Director of the Oak Ridge Reactor Safety Program said in 1971 that in view of these experimental studies of flow blockage he doubted that the emergency cooling systems would perform their intended function.[28] Detailed knowledge of flow blockage effects is missing and further experimental studies will be required before it can be resolved.

A third difficulty with ECCS effectiveness was identified in the spring of 1972 when spent fuel was unloaded from a power reactor in Rochester,[29] New York. It was found that unexpected shrinkage of the uranium oxide fuel pellets had occurred. The shrinkage allowed the pellets to shift downward within the fuel rod cladding. The motion left random gaps between pellets up to 3 inches in length. The reactor operating pressure had been large enough to overcome internal gas pressures within fuel rods and had crushed and collapsed the clad at many of the gaps. It was soon realized that this fuel damage would severely aggravate a major accident. Consideration of the phenomenon increased predicted maximum accident temperatures by as much as 600°F.[30] It was apparent that it could possibly ensure loss of accident control.

The problem is serious also because it represents an anomalous development in the one critical part of reactor design that is said to have received more study than any

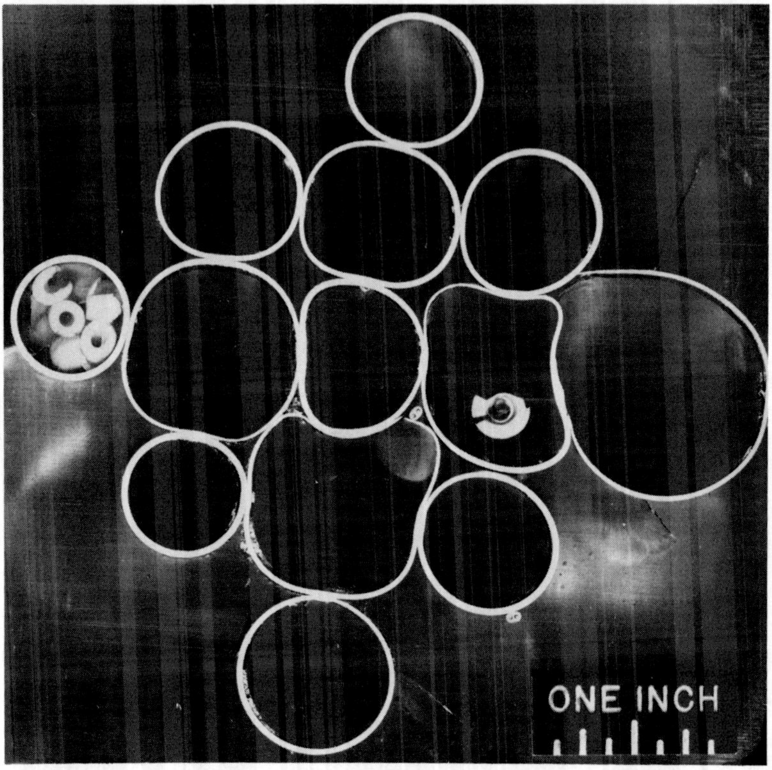

Figure 4.11 Cross section of an experimental fuel bundle showing the swelling that occurs during a loss-of-coolant accident. The blockage can prevent the flow of emergency cooling water and thus contribute to a core meltdown. Photo from ORNL-4752, Jan. 1972. Used by permission.

other: behavior of the uranium fuel. All present reactor fuel is now believed susceptible to such shrinkage. In consequence, power cutbacks and fuel management changes in a large number of both pressurized- and boiling-water reactors have been required in order to increase the margins of safety.

4.16 Lack of Conservatism in Computer Codes

An attempt by industry and the AEC has been made to incorporate conservatism in the computer analyses of ECCS performance. This is a standard practice employed to accommodate uncertainties in safety system analyses. The purpose is to avoid overestimating safety margins, or, to put it another way, avoid underestimating accident severity. It is of great importance to know the degree of conservatism so that one can be assured that uncertainties in a large number of complicated phenomena are adequately accommodated. Much is made of the fact that conservatism is added to the computer model to cover possible nonconservatisms in the treatment of some phenomena. Unfortunately, it is not now possible to establish the overall conservatisms in the present analyses and it is only recently that the AEC has started to address the problem of determining the uncertainties.

Recent history has shown a continuous erosion of claimed margins of safety, one of the principal examples being the decreasing expected reflooding rates. The AEC, in 1972, in reviewing research results of the previous year, frankly stated that there was "an observable trend" such that "new experimental data and additional calculational results have in several cases suggested the need for additional conservatism." The official AEC view, and the opinion of the reactor manufacturers, is nevertheless that there is sufficient conservatism in the safety analyses. In spite of the many matters remaining unresolved, the AEC believes there is "reasonable assurance" that the ECCS <u>will</u> work to bring the consequences of a pipe rupture confidently under control.

It appears to us in viewing all of the available information that assurance of adequate ECCS operation has not been achieved. Moreover, we believe there are certainly some credible accidents in which the emergency systems will surely fail to prevent core meltdown. In any event, and what is most important, the possibility is not excluded on the basis of present evidence. The former Director of the AEC's Division of Reactor Development and Technology said in an internal memorandum[31] in 1971: "No assurance is yet available that emergency coolant can be delivered at the rates intended and in the time period prior to clad and subsequent fuel melting due to decay heat generation."

4.17 Problems in Design, Construction, and Operation

In addition to the difficulties in assuring that loss-of-coolant accidents can be controlled, a number of other troubles have surfaced in the design, construction, and operation of reactors in the nuclear power program. Defects in design and in construction may signal other weaknesses, present but unidentified; and accidents and abnormal operating experiences that have occurred suggest ways that such weaknesses may, combined with human error, aggravate otherwise innocuous events. None of the many accidents that have occurred has developed to large scale or involved the release of radioactivity in damaging amounts. Even so, the record is disquieting and especially so since the total operating experience with the large reactors is quite small. Only about 30 reactors have been generating electricity for over one year, and only about 60 reactor years of experience has been gained with the larger units.

There has been a broad breakdown of quality assurance in the reactor program. Quality assurance in specification, fabrication, installation, and maintenance of plant components is relied on in an important way to minimize the likelihood of reactor accidents. However, the AEC has concluded[32] that fully effective quality assurance programs "have not been sufficiently developed or implemented," evidently based on their finding that "the number of defects, equipment malfunctions, or failure events that have been encountered during [reactor] construction, pre-operation testing, and routine nuclear operation has been large...." Some accidents have occurred, the AEC states, in which "the absence of more serious effects was largely the result of good luck."[33]

4.18 Examples of Operator Error and Plant Malfunction

One pressurized water reactor vessel was subjected to more than twice its permissible maximum pressure when a plant operator was called to the telephone and left a reactor charging pump unattended. The reactor fortunately was far below normal operating temperature at the time of the event.

The integrity and operability of valves in nuclear power plants has been of continuing operational concern. Indeed an AEC study stated that it appears "highly possible that the malfunctions of valves actually is [sic] increasing."[34] In one plant, 150 valves had faulty valve-operator switches, of which 10 failed, with similar failures in three other plants. A malfunction of a 10-foot valve at a large boiling-water reactor caused the valve to slam shut and induce a water surge in a steam condenser cooling water line. The surge, or water hammer, ruptured an expansion joint causing flooding of the turbine building basement to a depth of over 15 feet with river water. The flood disabled 19 pumps, emergency diesel power supplies, and <u>all</u> of the emergency core-cooling systems. The reactor was shut down at the time of the incident; otherwise events might have been far more serious.

An accident at another operating boiling water reactor was initiated by operator error. It was aggravated by the failure of a primary coolant relief valve. Loss of 50,000 gallons of reactor coolant to the containment structure ensued. Two further and important valves in the steam system failed in the course of the accident and, as a consequence, one isolation condenser, a plant protective feature, was unable to function.

In an 18-month period in 1971-1972, there were 121 valve malfunctions or failures, many of them multiple-valve events involving a total of 250 valves in 22 reactors. In a later 5-month period, there were 81 such events. In early 1970 several 28-inch valves in the primary coolant systems at boiling water reactors were found to have walls thinner than specifications allowed. Subsequent studies revealed similar difficulties at 20 different reactors, of both pressurized water and boiling water types. Of some 633 valves studied, over 15% had excessively thin walls. Some installations reported that documentation was insufficient to establish whether or not their valves met required specifications. So far, valves with substandard walls have not caused any reactor accidents.

A heavy rain at the Turkey Point reactor near Miami, combined with a plugged drain, caused plant flooding of such proportions that safe operation of the unit was threatened.

Three ruptures have occurred in pressurized water reactor steam systems. These systems are built to the same quality specifications as primary coolant lines. A number of workmen were injured and two were killed in these accidents and considerable damage done to installations. One of the accidents resulted from such underdesign that components failed to withstand normal operating temperature and pressure. Another of the ruptures was apparently the consequence of an inadequate weld. The cause of the third, complete rupture of a steam pipe, was never identified.

In addition to these ruptures, a 180° crack recently occurred in an 18-inch secondary coolant line inside the containment in an operating pressurized-water reactor near New York City. The escaping steam impinged on the 3/8-inch thick inner steel liner of the containment structure and caused unexpected buckling of the steel over a 40-foot span. This has raised the wholly new issue of whether or not containment integrity would persist through the course of a loss-of-coolant accident even if the accident were otherwise controlled. The matter has not been resolved.

Substandard and defective welding has plagued the nuclear industry. Continued complaints by a welding inspector at one reactor ultimately resulted in the formation of a review committee which verified the existence of 94 defective welds of 650 inspected at the installation. A similar review at one of a pair of operating reactors found defective 4% of nearly 5000 welds. About 1000 of the welds were unavailable for inspection and the review concluded that at least the same fraction of the remainder were presumed defective.

One of the important boiling water reactor emergency core-cooling systems, the high pressure coolant injection system, has failed, in several reactors, to come into service over 20% of the time in tests. When called on 5 times during reactor operations, the system failed to function 3 times. In these particular instances the systems were not required for the mitigation of potentially serious accidents. There are several dozen occurrences in which the very important diesel emergency power sources have failed from such mundane causes as rusty components and defective maintenance. At one midwestern reactor, part of the reactor core-support structure was caused to fail by unexpected vibration, leaving the entire core free to oscillate and to wear and damage its supports and shift in position. Excess vibration of core components in yet another reactor has caused extensive damage to fuel assemblies, including cracks up to 1½ feet long. Similar damage was discovered in another reactor. Cracks have been found in important reactor piping, in one case 1¼-inch-long cracks in primary reactor coolant lines.

One series of important accidents in boiling water reactors, nearly eighteen in number,[35] has occurred because of an unexpected and generic difficulty in controlling the reactor water level under abnormal circumstances.

Loss of water level control in these accidents has caused unscheduled pressure-vessel blowdown and partial loss of coolant. The accidents have led to damaging exposures of containment instrumentation and control equipment to high temperature steam. A wide variety of different events, many of them minor, have initiated unstable behavior which has in turn resulted in more serious consequences. Reactor coolant loss in the events has been substantial, frequently in the tens of thousands of gallons, although in no case so far has a core itself been left entirely without cooling water. The accidents have been characterized by aggravation, occasionally severe, arising from other unsuspected design errors, component failures, problems with instrumentation, operator error, and maintenance- and operating-specifications defects. These accidents have generated considerable concern not only regarding their implications but also because of the damage they have caused - in one case over $16 million. One of the design assumptions of boiling water reactors was that the release of many tens of thousands of gallons of superheated primary coolant to the containment was never intended except in "rare emergencies to prevent primary system overpressurization." Such events, however, have been occurring at the rate of one each two reactor-years, an unpleasant surprise.

Recently an anonymous letter informed the AEC that the main steam lines in two nearly-completed reactors were situated so that rupture could disable the entire residual-core-heat removal system and thus likely result in core meltdown. The steam lines also passed close to the reactor control rooms where the rupture could have killed the operators and destroyed the control instrumentation, also with the risk of core meltdown. These defects had entirely escaped notice in the design and regulatory reviews associated with construction of the reactors. The cost of shutdown and repairing of the defect in one of the two plants exceeded $20 million. Steam line placement difficulties have since been identified in nearly 20 reactors. In addition, several similar cases have been identified involving unsatisfactory orientation of steam turbines.

4.19 The AEC's Internal Review of Reactor Operations

In 1973 the AEC initiated a task force study of the reactor licensing process, directed by the AEC's assistant director of regulation. It was completed in October 1973. The existence of the study was not made public nor the results released by the AEC. A copy of the report received by the present authors was made public in early 1974. The task force report spoke of "the history of problems besieging reactors under construction and in operation..." and said "there is still an unanswered question as to the quantified degree of safety (or, conversely, the level of risk) of a nuclear power plant." The task force "does not believe" that there is "the required confidence level that the probability for such a [major] accident is one in a million or less per reactor-year."

The report continued:

"Approximately 850 abnormal occurrences were reported to the AEC during the 17 month period used as a sample base (January 1, 1972 to

May 31, 1973). These abnormal occurrences involved malfunctions or deficiencies associated with safety related equipment. Forty percent of the occurrences were traceable in some extent to possible design and/or fabrication related deficiencies. The primary cause of at least 200 of the component malfunctions was design and/or fabrication errors. The remaining incidents were precipitated by operator error, improper maintenance, inadequate erection control, administrative deficiencies, random failure, and variations of the foregoing. Many of the incidents had broad generic applicability and potentially significant consequences..."

The task force conclusion concerning this accident record was that

"The large number of reactor incidents, coupled with the fact that many of them had real safety significance, were generic in nature, and were not identified during the normal design, fabrication, erection, and preoperational testing phases, raises a serious question regarding the current review and inspection practices both on the part of the nuclear industry and the AEC. This is particularly true when the increasing number of operating reactors which will be on-line in the 1980's and 1990's is considered."

The Rand Corporation, following a review of reactor accidents, noted the "...increasing reports of poor quality control and documented carelessness in manufacture, operation and maintenance of these complex machines."[36] This record, coupled with the major weaknesses in ECCS capacity, has led the present authors to conclude that a reactor accident with major consequences is a real possibility in the coming decades. Walter Jordan, formerly assistant director of Oak Ridge National Laboratory, believes that: "Scientists and the public should be prepared to face the possibility of a nuclear [reactor] incident just as we expect major earthquakes that will exact a large toll in property and lives."[37]

References

1 C.G. Lawson, "Emergency Core Cooling Systems for Light-Water-Cooled Power Reactors," Oak Ridge National Laboratory (ORNL-NSIC-24), Oak Ridge, Tenn. 1968, p. 52, and the report of the task force established by the USAEC to study fuel cooling systems of nuclear power plants: "Emergency Core Cooling," 1967, p. 145.

2 "The Safety of Nuclear Power Reactors and Related Facilities" (WASH-1250) USAEC (June 1973). There are substantial criticisms of this report. See especially "Nuclear Safety: AEC Report Makes the Best of It," R. Gillette, _Science_, Vol. 179 (Jan. 26, 1973).

3 "Water Reactor Safety Program Plan" (WASH-1146) USAEC (February, 1970).

4 Nuclear Safety," four articles by R. Gillette, _Science_, Sept. 1, 8, 15, and 22, 1972.

5 _Federal Register_, _36_ 12247 and 12250.

6 D.F. Ford, H.W. Kendall, I.A. Forbes, and J.J. MacKenzie, "Cooling Water," _Environment_ _14_, No. 1 (Jan./Feb. 1972) and "Nuclear Safety", _Environment_ _14_, No. 7 (Sept. 1972).

7 USAEC Emergency Core Cooling Systems Hearings, AEC Docket RM-50-1.

8 "An Assessment of the ECCS Rulemaking Hearings," prepared by the Union of Concerned Scientists (UCS). Available from UCS, Cambridge, Mass., and Friends of the Earth, San Francisco, Calif.

9 "Loss of Coolant/Emergency Core Cooling Augmented Program Plan," Aerojet Nuclear Co. (August 1971). Available as Exhibit 1039, ECCS Hearing Record.

10 W.B. Cottrell, then-Director, ORNL Reactor Safety Program, letter to L.M. Muntzing, Director of Regulations, USAEC (Dec. 6, 1971). Available as Exhibit 1020, ECCS Hearing Record.

11 See "Nuclear Safety" in reference 6.

12 "Theoretical Possibilities and Consequences of Major Accidents in Large Nuclear Power Plants" (WASH-740) USAEC (Mar. 1957).

13 "Nuclear Reactor Safety: An Evaluation of New Evidence," UCS, Cambridge, Mass. (July 1971), reprinted in _Nuclear News_ (Sept. 1971).

14 AEC Internal Files on 1964-65 Update of WASH-740. Released June 1973.

15 Peter Morris of USAEC, Directorate of Regulatory Operations, at Meeting of International Atomic Energy Agency, Julich, Austria (Feb. 5-9, 1973).

16 _First Report from the Select Committee on Science and Technology_, "The Choice of a Reactor System," British Parliament, Session 1973-74 (Jan. 29, 1974).

17 "Fail Safe Systems for Reactors," Sir Alan Cottrell, FRS, letter, _Financial Times_ (London) June 7, 1974.

18 Announcement, House of Commons, Britain.

Reported in Wall Street Journal (July 11, 1974).

19 Battelle Memorial Institute Reports, BMI-1825 and BMI-1910 (Columbus, Ohio).

20 George Brockett, "Interim Draft: A Review of Analysis Methods for Evaluating ECCS Performance During a Loss-of-Coolant Accident." Part II, BWR's (April 1971), Part III, PWR's (April 1971). Available as Exhibits 1033 and 1039, ECCS Hearing Record.

21 AEC Internal Memorandum from M. Rosen and R. Colmar to ECCS Task Force (June 1, 1971). Available as Exhibit 715, ECCS Hearing Record.

22 "Water Reactor Safety Program Augmentation Plan," USAEC, Division of Reactor Development and Technology (Nov. 1971). Available as Exhibit 1026, ECCS Hearing Record.

23 Alvin Weinberg, letter to then AEC Chairman James Schlesinger (Feb. 1972).

24 See note 8.

25 ACRS Letter to AEC Chairwoman D. L. Ray, Nov. 20, 1974, "Report on Water Reactor Safety Research."

26 See note 8.

27 D.F. Ford and H.W. Kendall, "An Evaluation of Nuclear Reactor safety," Vol. I (Mar. 1972), the UCS Direct Testimony, Exhibit 1041, in the ECCS hearings. This reference contains extensive analysis of the problem.

28 Minutes of Meeting, ORNL Nuclear Safety Program Annual Information Meeting, A.A. Katterhenry (Feb. 16-17, 1971).

29 See note 8 and R. Gillette, "Nuclear Safety: Damaged Fuel Ignites a New Debate in AEC," Science 177 (July 28, 1972).

30 James S. Moore, Testimony, p. 15,297, ECCS Hearing Record.

31 M. Shaw, then-Director of USAEC Division of Reactor Development and Technology, letter to R.E. Hollingsworth, General Manager, USAEC (Feb. 17, 1971). Available as Exhibit 1112, ECCS Hearing Record.

32 See note 2 and also Remarks, W.O. Doub, Commissioner USAEC, released by USAEC (Nov. 29, 1973).

33 USAEC, "Reactor Operating Experiences," ROE 69-9 (1969).

34 USAEC, "Reactor Operating Experiences - Valve Malfunctions in Nuclear Power Plants" (Undated).

35 Studies of a portion of these accidents have been analyzed in an AEC volume: "Evaluations of Incidents of Primary Coolant Release from Operating Boiling Water Reactors," USAEC Directorate of Regulatory Operations (Oct. 30, 1972).

36 "California's Electric Quandary," R. Doctor et al., Rand Corporation, Santa Monica, Calif. (Sept. 1972).

37 Walter Jordan, *Physics Today* May 1970.

Chapter 5

THE NUCLEAR SAFEGUARDS PROBLEM

Lawrence Scheinman

Nuclear power presents an added dimension to the issue of public health and safety, one that arises from the fact that the same nuclear material that is capable of providing light for a city's thoroughfares, buildings, and residences is, when fabricated correctly and employed as a weapon, also capable of obliterating that city and its inhabitants. The task of minimizing the risk that nuclear material will be diverted to destructive ends is referred to as the nuclear safeguards problem. It is the subject of this chapter.

5.1 Nuclear Weapons Production[1]

Nuclear weapons can be constructed from uranium-235, which occurs in nature. Natural uranium is composed of 0.7% of this fissionable material, plus 99.3% of nonfissionable uranium-238. Natural uranium is processed to provide reactor fuel containing 3% uranium-235 and must be further processed to 90% uranium-235 for the construction of weapons. Presently developed separation procedures, referred to as enrichment, require a massive technology far beyond the capacity of most nations.

Nuclear weapons may also be constructed from plutonium-239. This material does not occur naturally, but is produced during normal power reactor operation from nuclear reactions involving the 97% of the uranium fuel which is otherwise nonfissionable. Plutonium is separated by chemical techniques from partially or completely spent fuel. Spent fuel is extremely radioactive from the "ashes" or fragments of fissioned U-235.

Plutonium is accumulating from the U.S. nuclear power program and may be utilized in place of U-235 as a reactor fuel. The United States is already making plans to recycle the stores of plutonium and, indeed, some is already in place in at least one U.S. reactor. By the early 1980s it is likely that much reactor fuel will contain plutonium. Chemical separation of plutonium from fresh fuel is far easier than from spent fuel because fresh fuel is only weakly radioactive and requires less complex separation procedures.

A simple implosion nuclear weapon requires from 4-8 kilograms of plutonium, depending on the weapon design. A typical large reactor produces in excess of 200 kilograms of plutonium per year. Accordingly, a few months' operation of such a reactor would produce enough material for a number of weapons. Continued operation of a reactor without separation of the produced plutonium slowly builds contaminating isotopes of plutonium that cannot be separated chemically from plutonium-239 and that make the construction of reliable nuclear weapons less certain. Nevertheless, plutonium, even from spent fuel, can be used in the construction of explosive devices.[2]

5.2 The International Safeguards Problem

In the decade following the devastation of Hiroshima and Nagasaki in 1945, the United States pursued a policy of nuclear secrecy in the hope that access to nuclear technology could be confined to the fewest possible states unless and until such time as a system could be devised to control atomic energy. These convictions were reflected in the Baruch Plan which called for the creation of an International Atomic Development Authority to own, control and operate facilities producing fissionable materials.

The Baruch Plan failed to gain the necessary support in the international community. Nuclear technology, however, began to spread and the promise of a peaceful atom gained in credibility. As it became more and more evident that technological spread could not be prevented, and that the technical knowledge associated with nuclear energy was no longer the province of the select few, the United States took the initiative in December 1953 of inaugurating the Atoms-for-Peace plant which entailed, among other things, a program for the controlled dissemination of nuclear technology and material and the creation of an International Atomic Energy Agency (IAEA) to facilitate and regulate this task.

Controlled dissemination meant safeguards, not of the very comprehensive type envisioned in the Baruch Plan, but nevertheless of a type adequate to reduce to a minimum the risks to international security inherent in the transmission of nuclear material and know-how. Although the character and expectations of safeguards have modified over the years, it would be appropriate to define safeguards as the body of techniques and measures designed (1) to guard against the diversion of nuclear materials from legal or permitted uses, and (2) to provide timely detection of such diversions from those specified uses as might occur. The underlying hope of advocates of safeguard systems is that the threat of detection will serve to deter potential divertors.

Most analysts of the history of international safeguards would agree that there has been a retreat from the principle of comprehensiveness. Most would also agree that the reason for this retreat lies in the fact that international safeguards involve intrusion on national sovereignty. Such intrusion is not lightly regarded by the states whose consent is necessary to the creation and implementation of an international safeguard system. National systems of self-inspection would be more readily acceptable to nation-states and have been increasingly recognized as essential components of a good material-management system. Purely national systems, however, would not be given high marks for credibility by other states whose interests demand assurance that there is no diversion of materials intended for peaceful purposes to military programs whose completion might one day threaten the security of neighboring states.

The temporary solution to this problem has been the adoption, in the context of the Non-Proliferation Treaty (NPT) of 1968, of a safeguard system whose scope and comprehensiveness, it is hoped, will strike a reasonable and credible balance between sovereignty and the minimal requirements of an effective internationally anchored control system. The anchor is the International Atomic Energy Agency which has had similar but substantially more modest safeguard responsibilities during the past decade.

An important feature of the IAEA/NPT safeguard system now in effect is that rather than establishing a self-sustaining safeguard operation, it relies on national control systems whose integrity and reliability the international agency will verify through procedures involving accounting, surveillance, and inspection. On the principle that any system is only as strong as its weakest component, we must conclude that the ultimate effectiveness of the NPT safeguards system will depend on the sophistication, good will, and understanding of the participant nation-states.

The preceding discussion has addressed the dominant concern of safeguards systems to date: the conditions under which nuclear material can be disseminated without increasing risks to international security, and the ways in which international institutions can be brought to bear on the problem of national diversion. That is to say, the basic concern of the principal nuclear supplier states (United States, Canada, Great Britain) has been how to prevent or inhibit the horizontal spread of nuclear weaponry while at the same time not denying to non-nuclear-weapon states the benefits of the benign atom, a denial which in any event could only have been effective for a limited time.

5.3 The U.S. Diversion Problem

The relevance of the problem of national diversion is in no way diminished if we turn attention to another serious matter: the risk of diversion or theft of nuclear material within a country that already has acquired a nuclear weapons aresenal. This is the problem in the United States today. It is an issue that only recently has received the serious attention it deserves. The heightened sensitivity to the safeguard and physical security risks associated with a nuclear economy is a function of at least three things: First, there is the fact that large quantities of <u>special</u> <u>nuclear</u> <u>material</u> (material from which weapons may be constructed: U-235, U-233, Pu) now are or shortly will be flowing through the nuclear fuel cycle, thus significantly increasing the opportunity for diversion or theft. Each new power reactor to go on stream adds to the potential plutonium reserves of the country. Second, the new awareness is fed by the rise of dramatic and occasionally suicidal acts carried out by terrorist and revolutionary groups who seek to win their objectives at virtually any price and by a concern lest such groups seize upon nuclear materials to implement their threats. In the third place, there is recognition that a commodity of such high value as plutonium ($9000/kg) or enriched uranium ($20,000/kg for fully enriched uranium) will eventually become an irresistible target for criminal organizations bent on gleaning the profits that the sale of such material would bring in a nuclear black market. The possibility for the development of such a market must not be underestimated. Its clientele could range from the well-to-do eccentric who might want to change the world to avoid - or bring about - Armageddon, through the terrorist groups mentioned above, to the non-nuclear-weapon state that has ratified the Non-Proliferation Treaty. Such a state might not wish to risk getting caught violating its obligations not to divert nuclear material in its fuel cycle from peaceful to military purposes, yet want to acquire clandestine nuclear material for security or other reasons.

The potential weaknesses of the nuclear

fuel cycle are important insofar as they suggest certain forms of protection and counteraction. But the collective problem of the availability of large quantities of extremely dangerous materials in our society rises above any specific problem and suggests the general question of whether, and under what circumstances, we can afford the luxury of a nuclear civilization.

The technical knowledge necessary to fashion a crude nuclear weapon can be derived from unclassified literature such as Samuel Glasstone's Sourcebook on Atomic Energy, or the Los Alamos Primer published by the Atomic Energy Commission. As the leading nuclear weapons designer Dr. Theodore B. Taylor has pointed out, there are some problems associated with weapons fabrication but they are not at all insurmountable. Highly enriched uranium in a wide variety of forms can serve as the base for a nuclear weapon, and the same is essentially true with respect to plutonium. Both materials are shipped around the United States between fuel fabrication, reprocessing, and storage facilities in usable or convertible-to-use form as a metal, oxide, or - in the case of enriched uranium - as uranium hexafluoride, whose properties render it easily reducible to metallic form.

The amount of fissionable material necessary to fabricate a weapon (i.e., to develop a critical mass) powerful enough to threaten a small city or a large segment of the population of a metropolitan area (likely targets for terrorist activities) is very small indeed - approximately 20 kilograms of fully enriched uranium or 5 kilograms of plutonium.[3] Due to the high density of these materials the amounts in question are no larger than the size of a softball. Furthermore, even lesser amounts of nuclear material could be used as a sort of radiological contamination device not involving an explosive weapon but with toxic capabilities of enormous proportions.

5.4 The Availability of Special Nuclear Materials

The chief obstacle a group faces in making a nuclear weapon is obtaining the fissionable material. Uranium-235 can be separated to the required concentration from naturally occurring uranium only with the greatest difficulty, and plutonium is produced only in nuclear reactors. In obtaining weapons-grade material, then, diversion of someone else's supply can be a very attractive, inexpensive means. While absolute prevention of thefts is not possible, the important questions are whether current physical protection systems are adequate to deter would-be thieves, or if not, whether there is a reasonable chance of preventing the successful seizure of nuclear material. The prevailing responses at this time, evidently, are no.

An assessment of the impact of the fuel cycle must consider diversion of strategic nuclear materials because of both the projected growth in the amount of such materials in the society and the increasing ease with which they may be fabricated into nuclear weapons as the technical problems and their solutions become more widely known.

The projections for the period to the end of this decade indicate that we will be accumulating plutonium in amounts on the order of 10-25,000 kilograms per year. A very small quantity of this diverted to the black market would be adequate to allow terrorist activities. The aggravation of this situation, if and when breeder reactors are in active use, is very great because <u>the breeder uses weapons-grade material in large amounts</u>. This increases the amount of plutonium in the system that has to be transported. The weaknesses in the transportation system have, like the consequences of diversion, been noted by a number of experts.

In a paper presented in November 1972 at the International Conference of the Atomic Industrial Forum and the American Nuclear Society, Mason Willrich from the University of Virginia Law School stated:

"The U.S. national safeguards system, which should include effective measures for physical protection and recovery as well as accountability, is still primitive. Under existing regulations, for example, it is possible that a large amount of special nuclear material may be stored in a locked building that is checked by an unarmed watchman only once every four hours. In the shipment of a large amount of special nuclear material by truck, the driver is not required to be accompanied by someone else, nor armed, nor even in continuous communication with any authority. Moreover, there are no regulations in effect which specify procedures for the recovery of stolen material."

He went on to say:

"Lacking an overall rationale, the U.S. national safeguards system is embodied in a patchwork of AEC regulations and (ad hoc) arrangements with particular licenses and contractors. Lacking a single focus of responsibility for policy, planning and control activities aimed at preventing nuclear diversion, it's doubtful that the Commission can either develop or implement a systematic approach to the problem. The AEC cannot expect industry to cope effectively with its side of the nuclear material control problem unless it specifies clear cut objectives and detailed procedures in comprehensive regulations."

5.5 Weaknesses in Transportation

It is apparent then that we have a critical and important problem facing us. Unhappily the record of effectiveness in safeguarding these materials in our nation has so far not been entirely satisfactory. Considerable effort and some resources have been devoted by the AEC to the question of safeguarding strategic materials while they are in the processing plants, by methods of accountability, inventory, development of analytical and assay techniques, and other techniques of this kind. However, there has not been an equivalent emphasis on weaknesses in the transport of these hazardous elements. Special nuclear materials for military purposes are treated differently in transport from materials for peaceful purposes, and the guarding of shipments designed for peaceful purposes is much less stringent and much less effective than for the military case. Indeed, Brigadier General D.L. Crowson, former head of the AEC's Office of Safeguards and Materials Management, noted that the strategic nuclear materials destined for civilian purposes are usually shipped like "a special delivery letter." There have been a number of examples where quantities of materials equivalent to a number of nuclear weapons have been inadvertently diverted in transit and temporarily lost. There have been other cases noted in which plutonium has apparently disappeared permanently. A recent article in Science magazine pointed out that, in 1969, the SEFOR reactor was deficient in "a few kilograms of plutonium"; and the Nuclear Materials Enrichment Corporation in Pennsylvania noted that, for over six years, 6% of the strategic material it handled was unaccounted for.[4]

The transportation environment appears particularly susceptible to loss or theft. Conventional wisdom and national mythology tell us that the hijacking of trucks carrying valuable merchandise is part of the American landscape; that organized crime is invariably identified with the trucking industry; and that many a shipper knows that what is not stolen has a good chance of being lost. Added to this in recent years is the spate of skyjacking incidents to which a dramatic cast has been given in view of occasional allegations that skyjacked aircraft have been carrying high-grade special nuclear materials.

Whether the allegations are true or false (AEC regulations recently were amended to sharply curtail the quantities of special nuclear material that could be carried aboard passenger aircraft) is less important than the question of whether adequate attention has been given to the vulnerability of nuclear material in the transportation phase of the nuclear fuel cycle. The facts simply are that the different components of the fuel cycle are geographically dispersed, thus requiring a substantial amount of physical movement of nuclear fuels in various forms - as fabricated fuel components, as spent fuel elements, as reprocessed material, and as ores or concentrates.

The problems identified with the transportation of nuclear fuels relate to other phases of the fuel cycle as well. A matter of some concern is the adequacy of regulations governing nuclear fuel storage. Prior to recent amendments to 10 CFR 73 (the AEC regulations issued in 1969 to cover use, storage and transportation of strategically important quantities of nuclear material) physical security provisions for material not in transit required that such material be placed in an area enclosed by physical barriers to which access was controlled by authorized persons. Surveillance was to be either by an intrusion alarm system or by periodic checks by armed or unarmed guards.

These inadequate regulations have been revised to provide for more intensive control of special nuclear material. It is difficult to say to what extent the new regulations truly augment physical security of special nuclear material: licensees are required to establish materials access areas consisting of physical barriers, controlled access and egress, and detection and surveillance equipment and procedures. Measured against the array of skills and resources that determined divertors or thieves might bring to bear, the question of adequacy remains open. Also unanswered is the question of just how much security and safeguards should be required in order to maintain a rational system.

5.6 New AEC Regulations

In November 1973 a series of amendments were adopted to earlier AEC regulations governing the physical protection of facilities handling plutonium and enriched uranium, the transportation of nuclear material and the frequency of checking physical inventories. The adoption of these new regulations has been regarded as corroboration of the criticisms of inadequate national safeguard provisions which have been made during the past year. Without passing judgment on the extent of AEC culpability, it would not be inappropriate to reemphasize our earlier point - that, as a general rule, in the past, safeguards have largely been directed toward coping with the problem of weapons proliferation. The problem of the proliferation of weapons simply loomed larger as a threat to international security than did subnational diversion or theft of nuclear materials. Additionally, until the mid-1960s, commercial nuclear power was more a hope than a reality and the magnitude of throughput in the nonmilitary segment of the nuclear fuel cycle was relatively inconsequential. This does not excuse the inadequacy of attention to domestic safeguard problems (a question of how much rather than whether attention was paid to the domestic program), but it does help to explain the rationale behind the priorities.

It would appear that the recent AEC amendments still do not go far enough in preserving and protecting public safety and security. Provisions for physical protection of special nuclear material in transit, for example, apply to shipments of more than 5000 grams of U-235 contained in uranium enriched to 20% or more. Whether the regulations (whose adequacy may be open to question) ought to be extended to low-enriched uranium, the dominant fuel now flowing in the nuclear fuel cycle, will undoubtedly be a point of future debate. For some, particularly industry, the ability of such material for weapons purposes is so low as to raise serious question as to whether the increased costs associated with more intensive safeguarding of such material is worthwhile. For others, the argument that even 3% to 5% enriched uranium has had about half the separative work necessary to achieve weapons grade material will prevail.

5.7 How Much is Enough?

The question of how much safeguard is enough, in a sense, captures the essence of much of the debate concerning appropriate levels of safeguarding: does a given added increment of physical protection warrant the cost associated with that increment? Should equivalent protection be applied to all nuclear materials (or even all special nuclear materials) at every stage of the fuel cycle, or should a safeguards system be graded in such a way that the character of the material, the nature of the facility, and the segment of the fuel cycle (transportation, reprocessing, storage, fabrication, etc.) govern and define the criteria to be applied in determining the adequacy of physical security provisions?

It has not been my purpose here to exhaust the range of problems associated with particular aspects of nuclear development, but rather to sketch the nature of those problems and to place them in the perspective of public choice concerning the role of nuclear power in national energy production. There can be no doubt in this regard that nuclear power presents problems of perhaps unprecedented magnitude.

Safeguards, as I have stressed, are tools applied to achieve certain aspects of this control. At the dawn of the nuclear age they were aimed at preventing diversion and misuse, but, in accommodation to the proliferation of nuclear technology and the political realities of the international system in which we live, emphasis gradually shifted from prevention to detection. In a sense, this may have been a necessary phase in the control of nuclear energy, for the accommodations made to political reality enhanced the general acceptability of the principle of safeguards. Thus, at the international level, the legitimacy of the safeguards concept is more firmly rooted than ever before.

At the national level, the legitimacy of the safeguards principle was never really in doubt. What has been in question is the scope and extent of safeguards requirements and the balancing of costs against those requirements.

There is apparently no projected increase in the money that is slated to go into these activities in the AEC in spite of the very great increase projected in the amount of materials which the system will have to deal with in the near future. Since there is evidence for inadequacies in the past, as we have noted, the situation can be expected to worsen unless higher priorities than at present are assigned to resolving the problems.

An additional aspect of the safeguards problem is that it has remained virtually completely outside the arena of public understanding and debate. It is clear that we must open up to the public the issues presented by this aspect of the nuclear economy. The whole nuclear debate has often been framed in technological language incomprehensible to the layman, or in gross generalizations which lead to the polarization of opinion without an adequate data base on which to make informed judgments. There is no reason why the issues cannot be posed in a comprehensible form so that information adequate to an informed judgment can be introduced. This is perhaps even more possible with respect to those issues that cluster under the rubric of "safeguards" and "public safety." At this juncture the questions are those related to the financial costs people are willing to bear in order to maintain given standards of living and life styles, and the risks they are willing to run as a result of the use, transportation, and storage of tremendous amounts of highly dangerous material.

5.8 Conclusions

It is my belief that nuclear safeguards, particularly in transportation, but elsewhere as well, have not had the priority that they deserve and are as a consequence weak and inadequate. They are apparently not projected to have the funding and resources which I believe they deserve; furthermore, although improvements have been mandated in the areas of transportation, plant protection (including armed guards) and materials accountability, it remains to be seen whether these will provide adequate security.

Thus there are choices to be made. I noted earlier, for example, the vulnerability of the transportation link in the nuclear fuel cycle. One alternative would be to establish a totally separate transport system for nuclear material possibly with assignment of transportation responsibility to federal authorities; another would be to establish a policy of concentration of nuclear fuel cycle facilities (reactors, reprocessing and fuel fabrication plants) in order to minimize the physical movement of nuclear fuels. Each alternative has its costs: in the former instance there would undoubtedly be an increase in cost which in turn leads to reevaluation of the economic competitivity of nuclear-produced power; in the latter case there would be environmental and safety factors associated with the siting of nuclear facilities, and the possible impracticality of such concentrated siting for economic reasons.

These issues cannot be resolved here. What seems essential, however, is that the costs (financial, security, health, environmental, etc.) associated with various patterns and rates of nuclear development be thoroughly explored and the results made available to the public for further discussion and debate. It should not be up to the AEC/nuclear industrial complex alone to determine what the tolerable and acceptable cost-benefit level should be. This should be a matter of public choice based on a thorough examination of the issues, the alternatives, and their consequences. We do not yet have a truly open forum in the nuclear field.

Safeguards, of course, represent only one piece of the nuclear puzzle, and one that has been more remote from the public consciousness than the problems of radiation hazards, nuclear accidents, or ecological risk. I have tried in this chapter to clarify the relationship of safeguards to a nuclear economy and to suggest some of the reasons for their need and some of the problems they reflect in the management and control of the development of nuclear technology, as well as some of the steps necessary for their resolution.

Attention to these matters must not be deferred. They deserve the kind and level of concern that is commensurate with the possible hazards presented by a successful diversion. In particular, we require vastly improved physical security for all transport, use, and storage. There has to be a substantial commitment to establish the risk and the probability of diversion and to select and develop the kind of improvements that are required.

Notes and References

1 This section is based on "The Proposed U.S.-Supported Egyptian and Israeli Nuclear Programs," by D.F. Ford, H.W. Kendall, G.B. Kistiakowsky, and G. Rathjens, Union of Concerned Scientists, Cambridge, Mass., June 25, 1974.

2 See, e.g., Victor Gilinsky, "Military Potential of Civilian Nuclear Power," in B. Boskey and M. Willrich (eds.) *Nuclear Proliferation: Prospects for Control* (New York, Dunellen, 1970), pp. 41-52; Mason Willrich and Theodore B. Taylor, *Nuclear Theft: Risks and Safeguards* (Ballinger, Cambridge, Mass., 1974), p. 12.

3 As Taylor further points out, appreciably smaller amounts of material may be required if the fabricator is sophisticated enough. Explosive nuclear devices can be miniaturized to a significant extent if the right mind and the right material are put together. See John McPhee, "The Binding Curve of Energy-II," *New Yorker* December 10, 1973, pp. 54,75.

4 Charles Thornton, former Director of Nuclear Materials Safeguards for the AEC, noted that "The aggregate material unaccounted for from the three diffusion plants alone is expressible in tons, no one knows where it is. None of it may have been stolen, but the balances don't close." Quoted in John McPhee, "The Curve of Binding Energy-1," *New Yorker* December 3, 1973, p. 126.

Bibliography

Ford, D.F.; Kendall H.W.; Kistiakowsky, G.B.; and Rathjens, G., "The Proposed U.S.-Supported Egyptian and Israeli Nuclear Programs," Union of Concerned Scientists, Cambridge, Mass., June 25, 1974.

Lapp, R.E. "The Ultimate Blackmail," New York Times Magazine, Feb. 4, 1973, p. 13.

Leachman, R.B., and Althoff, P. (eds.), Preventing Nuclear Theft: Guidelines for Industry and Government, Praeger, New York 1972.

McPhee, J. The Curve of Binding Energy. Farrar, Straus, and Giroux, New York, 1974.

Shapley, D. "Plutonium: Reactor Proliferation Threatens a Nuclear Black Market," Science 172 143-146 (1971).

Willrich, M.; Taylor, T.B., Nuclear Theft: Risks and Safeguards. Ballinger, Cambridge, Mass., 1974.

Chapter 6

THE TRANSPORTATION OF RADIOACTIVE WASTES: THE POSSIBILITY OF RELEASE OF CESIUM

Marc Ross

6.1 Introduction

Transportation of radioactive materials in connection with the operation of nuclear power plants can be divided into three categories:

1. transportation of fresh fuel to the reactor,
2. transportation of spent fuel from the reactor, and
3. transportation of other radioactive wastes from the reactor.

Typically, fresh fuel is brought from a fabrication plant by truck. The fresh fuel presents a potential transportation problem because it is very valuable and is thus, in principle, subject to theft and misuse. It is not, however, a particularly hazardous material under accident conditions. That is, the material would not diffuse widely into the environment and would not present a hazard to anyone approaching it. The annual needs for fresh fuel are about 30 metric tons or 6 truckloads for a 1000 Mw(e) reactor. A similar weight of spent fuel is carried to a reprocessing plant from the reactor in heavy casks, designed for that purpose, by rail or by truck. The spent fuel is particularly radioactive, generates considerable heat, and contains volatile and soluble components. Among these are several isotopes of cesium (Cs), fission products that are present in the spent fuel in considerable amounts. This chapter is concerned with the modes and consequences of the accidental release, during transportation, of the radioactive isotopes Cs-134 and Cs-137.

In addition to the spent fuel, nuclear reactor operation also generates other wastes. These latter are considered low-level wastes, and consist of sludges and resins arising from clean-up of the reactor and cooling system, as well as clothing, rags, paper, gloves and shoe coverings. All these low-level wastes are shipped from the reactor, either by truck or train, to low-level disposal facilities. About 2000 55-gallon drums of low-level waste are removed from each large reactor per year. The total radioactivity is about 2000 curies (Ci), and the heat generation is small.

6.2 Transportation of Spent Fuel

Of the three areas, the transportation of spent fuel offers the most serious hazard in case of accident or mishandling. I consider here the hazard associated with the release of cesium as a result of a hypothetical accident occurring during the transportation of spent reactor fuel. We shall see that the release of cesium presents a very grave problem, much more serious than the release of krypton and iodine considered in previous studies.

An accident in the transportation of light-water-reactor fuel to the reprocessing plant is assumed which results in loss of the cooling water from a cask. This eventuality has been previously considered,[1,2,3,4] but detailed attention was restricted to the gases krypton (Kr-85) and iodine (I-131). The loss of coolant will also lead to escape of some volatile and soluble fission products. The present study follows upon the discovery that cesium migrates in the fuel pellet during irradiation and substantial quantities escape and are deposited on the cladding. Furthermore, the cladding of the fuel rods in the postulated accident would experience much higher temperatures than in normal reactor operation. In this chapter, I estimate the amount of release of cesium from the fuel, from the fuel rod, and from the cask into the atmosphere, and find it to be substantial. I then consider the exposure to which a population might be subject at a distance from the cask. The isotopes involved, Cs-134 and Cs-137, are among the most potent known. The ensuing somatic effects and implied effects on land use are briefly discussed.

6.3 Quantity of Cesium Released from the Cask

One metric ton, MT, of spent reactor fuel after an irradiation of 33,000 megawatt-days per metric ton (MWD/MT) contains, after cooling off 90 days, 2.3×10^5 Ci of Cs-134, 1.1×10^5 Ci of Cs-137, 1.1×10^4 Ci of Kr-85, and 381 Ci of I-131, in addition to other radioactive materials.[5] Under typical licenses, the fuel must cool off at least 90 or 120 days, depending on the cask used. Spent fuel may be sent for reprocessing by road or rail. Typical large casks are water-filled, and contain 0.5 MT and 3.2 MT of spent fuel, respectively.[6]

An accident in transit could result in various degrees of damage to the cask leading to various modes and amounts of released radioactivity: (1) There could be zero release or no substantial release of cooling water (or radioactivity) in a period of, say, one day. (2) There could be seepage of coolant or flow of coolant through a small orifice or pressure crack such as might result from damage to a pressure release system, from distortion of the cask in the neighborhood of a gasket, or from excess heating of the coolant. Such seepage could result in substantial loss of cooling water over an hour to a few hours. This would normally result in drying out the cask as steam is released via some pressure release mechanism. (3) There could be a major breach leading to rapid loss of coolant.

It is difficult to estimate the probabilities of the various kinds of accidents or to characterize the impacts that would lead to the aforementioned release modes. The major cask-test standard involves impact against a flat surface. On the other hand impalement at substantial velocity, as may occur in rail accidents, or broadside impact at substantial velocity against curved objects, such as a bridge abutment, may be important. Accidents may involve multiple impact. In other words, there is a significant difference between test and design conditions, and real accident conditions. In addition, I stress that the quality of actual casks and the handling of these casks may not correspond to theoretically considered systems. Failure of workmanship, such as failure to secure bolts properly on loading, may result in substantial loss of cooling water in a minor accident or even in the absence of a road or rail accident. In the latter case the loss would in many cases be unobserved, especially in the case of rail transport.

An important possibility, not investigated here, is fire with or without associated impact. Fire could distort the cask or it could heat the coolant, causing its release. It could cause fuel rod rupture.[4] It could heat the spent fuel. Combustion gases could transport radioactive materials. To put this study in perspective, consider the typical prediction of 1000 light water reactors (LWRs) in the 1000-Mw region in operation at the turn of the century. This many reactors will imply about ten million miles per year of rail transport of large casks containing spent fuel, or, equivalently, 50 million miles of truck transport. Probabilities, on a per mile basis, of accidents at various velocities and with various times of duration of fire have been worked out and reviewed by the AEC.[7] In a ten-year period surrounding the year 2000, several rail accidents and/or several tens of road accidents involving casks would be expected with fires lasting well over one-half hour and/or impact at well over 30 miles per hour, as shown in Table 6.1. Considering this, as well as the possibility of faulty workmanship, I conclude that accidents resulting in damage of category (2) will not be unusual in a ten-year period.[8] Even the AEC considers the loss of cooling water in an accident in the design of water-cooled casks.[1,2,4]

One uncertainty in a transportation accident is the extent and degree of breakage of fuel rod cladding. (Here we are not interested in fuel rods which were perforated before transportation to the extent that the available cesium was left behind in the reactor core or the storage pool.) The cladding is brittle after irradiation. An impact leading to the breaching of the cask may impart considerable physical shock to the rods. It is not surprising, therefore, that authorities

Table 6.1 Expected number of serious accidents in transport of spent fuel casks in ten-year period surrounding the year 2000[a] (based on ref. cited in note 7).

Accident severity category	Vehicle speed at impact (mph)	Fire duration (hr)	Expected road accidents	Expected rail accidents
"Moderate"	0-30	1/2-1	150	8
	30-70	< 1/2		
"Severe"	0-50	> 1		
	30-70	1/2-1	4	0.2
	> 70	< 1/2		

a. Very severe and more moderate accidents have been omitted.

differ on this point.[2,4] I assume that 10% of the rods will be fractured at the time of the accident. (This number is not critical to the final result.)

In addition to possible damage to cladding upon impact and shortly afterwards, fuel rods which might otherwise remain intact are subject to creep rupture, i.e., slow and gradual breaking, at the high temperatures which develop after the cask has substantially dried out.[9]

Routes for cesium release associated with the two types of breach of cladding are shown in Figure 6.1. Of major interest is the amount of cesium which had already escaped

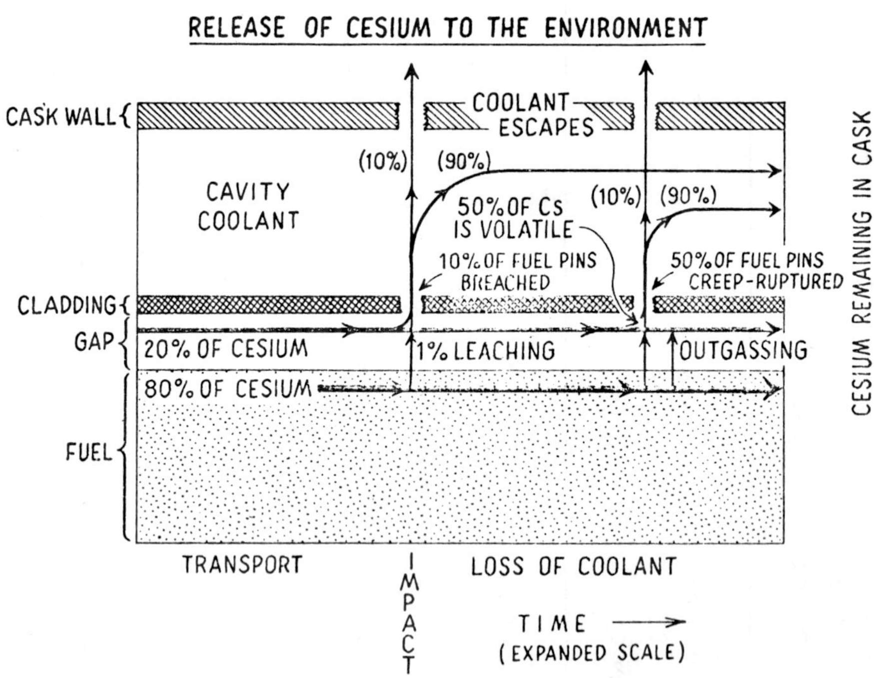

Figure 6.1 Routes of cesium release in an accidental rupture of a transportation cask.

from the fuel during irradiation. In typical light-water reactors, 30-35% of the noble gases escape from the interior of the fuel. Laboratory tests and calculations are the main source of information on escape of cesium. These can be correlated with rather limited direct evidence from actual reactor experience. The diffusion of cesium in sintered UO_2 is slow at the temperatures that occur at the outer curved surface of the pellet, but is comparable or greater than that of the noble gases in irradiated fuel at higher temperatures as shown in Figure 6.2.[10] Some cesium will diffuse out of the pellet; most will remain in a cylindrical distribution very near the outer curved surface. See Figure 6.3.[11] The migration of cesium to the surface shown

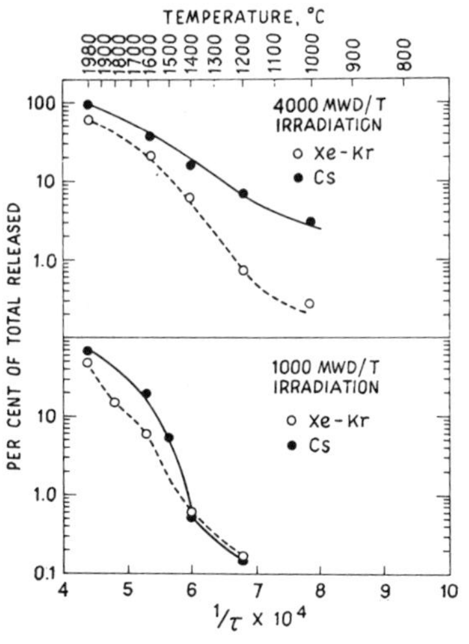

Figure 6.2 Release of fission products by diffusion from highly-irradiated PWR-type UO_2 heated 5.5 hours in purified flowing helium. Source: ref. cited in note 10. τ is the Kelvin temperature.

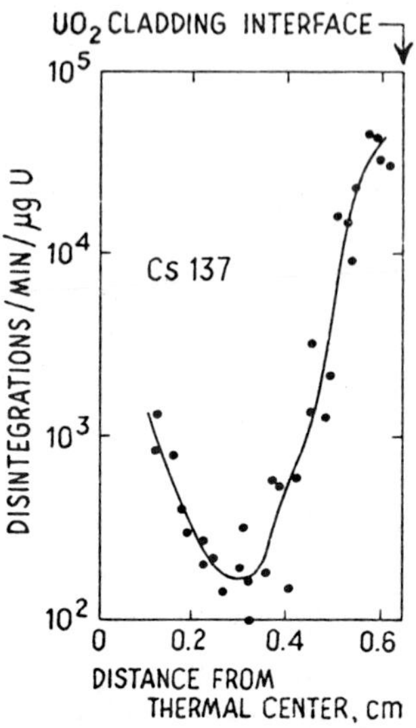

Figure 6.3 Radiochemical analysis of micro-drilled UO_2 sample, showing radial variation in cesium distribution at irradiation level of 1000 to 2000 MWD/MTU. Source: ref. cited in note 11.

in this figure would be even greater at higher, typical burn-up levels. In the report of one experiment it is stated: "The Cs-137 migrated both radially and axially and in the case of the highest-rated rod it is estimated that about 30% of the total rod inventory of Cs-137 is deposited on the cladding."[12] That is, the cesium which escapes from fuel pellets during irradiation condenses on the relatively cold pellet surface and inner wall of the cladding. The fraction escaping is sensitive to fuel temperature and thus to the location of the fuel rod in the reactor. I assume that 20% of the cesium produced in the fuel escapes from the fuel pellet into the gap between the pellet and cladding before transportation. My final result scales with this number.

A further source of cesium escape from fuel is leaching of the fuel in those rods which are breached at the time of impact. If the outer 1.2×10^{-5} inch of the fuel pellets is leached,[1,2] 0.01% of the pellet volume is involved, and we can conclude that about 1% of the cesium is leached on the basis of its high surface concentration. (This number is not critical to the final result.)

Under normal conditions in a large water-cooled cask the temperature of the cladding may be in the range 320-460°F.[13] After substantial loss of coolant, the temperature of the hottest rod will ultimately rise to 1250-1500°F depending on the cooling-off time of the fuel at the reactor and other considerations.[9] Specific calculations for the large irradiated-fuel cask, IF 300, indicate the even higher hottest-rod temperature of 1576° F.[4,13]

The fraction of intact rods which ultimately rupture under these conditions, and the time to rupture, is discussed in the Cask Designer's Guide.[9] If the cooling-off period has been short (90 days), most rods are expected to rupture. If it is longer (150 days), a substantial fraction of Zircaloy-clad rods are still expected to rupture within 10 hours of coolant loss. For the hypothetical accident to be discussed, we assume that 1/2 of all rods will creep rupture.[4]

6.4 Important Routes for Cesium Release

Two release modes must be considered: (1) For those rods breached at impact, cesium released from fuel during irradiation will dissolve in the cooling water and some more will be leached out of the pellet surfaces. Some of this cesium will be carried out of the cask with the steam and enter the atmosphere. (2) For those rods that creep rupture later, cesium released from the fuel during irradiation, if in volatile form, will enter the cask cavity as a gas or an aerosol. The metallic cesium vapor pressure is high at temperatures well below its boiling point of 1253°F at one atmosphere. Some of the free cesium in the cavity will be carried out of the cask with further pressure releases by the cask.

First consider a release from rods breached on impact. I have assumed above that 10% of the rods are breached and that 21% of the total cesium burden from these rods is taken up by the cooling water. As this water boils, a fraction of this cesium will be contained in particulates and escape from the cask and be carried off in the atmosphere as the steam escapes. Unlike quiescent vaporization, the bursting of bubbles results in an aerosol which has roughly the same content of solute as the body of liquid.[14] The fraction of cesium compounds entrained in the steam and escaping into the outer atmosphere is taken to be 10%. This may be a significant underestimate. The remaining 90% will be left on surfaces in the cavity. If we take all these factors into account, we arrive at an escape of about 2200 Ci of cesium into the atmosphere from a cask containing 3.2 MT of fuel.

Alternately, consider a delayed release after creep rupture of rods. I have assumed above that 50% of the rods creep rupture at high temperature, and that 20% of the total cesium burden from these has escaped the fuel pellets. I further assume that 1/2 of this available cesium is in volatile chemical form.

The chemical situation is complex and is the subject of current research.[15] It is probably safe to assume that 10% of the total cesium inventory of ruptured fuel rods goes into the cask cavity atmosphere as a gas or an aerosol. I do not explicitly consider further outgassing of the fuel as a source.

There are three sources of mass flow out of the cask at this time: boiling off of any remaining cooling water, ultimate increases in temperature, and the gases released from the ruptured rods. At the conditions under consideration, the partial pressure of the gaseous fission products is over 1/2 atmosphere. The mass flow which will carry some cesium out and into the atmosphere will depend on particular conditions such as cavity pressure and the pressure release mechanism at this stage. On the basis of the three sources of mass flow, I estimate that 10% of the cesium available in the cavity atmosphere will escape from the cask. The calculation using these estimates gives an escape of about 5300 Ci of cesium.

The overall result, then, of an accident involving the loss of cooling water from a 3.2 MT case of spent fuel, will be the likely release of 7500 Ci of cesium. (Roughly 2/3 of this activity will be due to Cs-134, half life, 2.3 yr; and 1/3 will be due to Cs-137, half life, 33 yr.) This release will take place over an interval of some hours.

6.5 Dispersal of Cesium and Its Somatic Effects

The problem of estimating the damage due to a release of cesium is difficult because it depends in detail on conditions which will differ widely from case to case. Rather than examining all kinds of weather conditions and all population distributions, I will examine two simple cases.

First, consider an individual located one kilometer downwind from the accident, remaining there while the radioactive cloud passes, or in a building with ventilation such that the exposure is the same. Consider stable weather conditions with a wind speed V = 1 m/sec (about 2 mi/hr) and a low degree of gustiness such that at 1 km there is a vertical spread σ_z = 8 m and a horizontal spread σ_y = 42 m.[16] Thus the cloud would sweep out a volume like a long narrow plume. Assume that the release is not carried high into the air by fire. (If it were, the second case below would be more applicable.) Assume that particulates are small and weather conditions are such that the cesium is not largely deposited out closer to the cask. Then if Q = 7500 Ci, the amount of radioactivity released in the accident, the concentration of the cloud downwind at 1 km will be

$$\frac{Q}{\pi \sigma_z \sigma_y V} = 7 \frac{\text{Ci-sec}}{\text{m}^3} .$$

I consider only the dose to an individual associated with inhalation. Breathing at a rate of 3.5×10^{-4} m^3/sec, an individual would have a total intake of 2.5×10^{-3} Ci during the cloud's passage. Assuming 75% uptake of cesium[17] and dose conversion of 0.86×10^{-4} millirem for each picocurie uptake (i.e., body burden)[18] of Cs-134 plus Cs-137 we find a dose commitment to the individual at one kilometer of 160 rem. Here I do not consider exposure subsequent to the passage of the cloud. However, it must be remembered that the radiation dose is not received all at once, but is governed by a biological half life in the individual of roughly 70-100 days.

At the dose level calculated, serious illness and a significant probability of fatality can be expected. A 300-rem sudden dose causes 50% fatalities.[19] The extended dose considered here would be much less lethal. Fetuses,[20] young children, and certain other people[21] would be most susceptible. The dose of 160 rem implies a dose of about 1 rem/day for some time.[17] This would significantly depress the bone-marrow and lymphatic systems, and elicit changes in immune responses. The probability of induced cancer within 25 years would be about 2%.[20]

The second case I consider is a windy, gusty day. Here it is most significant to consider the somatic effect of small doses of radiation suffered by people over a wide area and over a long period of time. Assuming the linear hypothesis, the BEIR report[20] concludes that the effect with respect to cancer of the sum of small doses averaging 0.1 rem/yr to the population of the United States would be roughly 3500 cancer deaths/yr (within 25 years).

According to analysis of Cs-137 distributions from weapons testing,[22] the dose commitment conversion factor is 1.44 mrem/mCi/km^2. That is, deposition at one time of Cs-137 at the level one mCi/km^2 results in a typical person living indefinitely in that environment acquiring a dose of 1.44 mrem from external decays. To more accurately assess the dose due to deposited Cs-137, factors for shielding by buildings and body tissue and for internal decays must be included.[22] The net result is a dose commitment factor of 0.67 mrem/mCi/km^2. The internal decay dose depends on particular food chains and might not apply at the site in question; about 50% of the consequent deaths quoted below are due to the external exposure alone. The dose from (1/3 Cs-137 plus 2/3 Cs-134) is less than for the same number of curies of Cs-137. Using an effective mean life in the environment of 14 years for Cs-137 for external irradiation,[23] it can be calculated that the Cs-134 - Cs-137 mixture is 0.80 times as effective per curie as Cs-137 alone. The dose in man-rems (a man-rem is a dose of one rem delivered to one man, or 1/2 rem delivered to each of two men, etc.) is thus $0.80 \times 0.76\ \rho Q = 0.54\ \rho Q$, where Q is the total cesium activity in curies and ρ is the average population per km^2.

Assuming a population of 1000 persons/km^2 (a typical figure for suburban populations) distributed over a very wide region downwind from the accident, the dose resulting from the assumed cesium release of 7500 Ci is 4.0×10^6 man-rem. Using the cancer induction rate from the BEIR report cited above, this long-lived contamination implies roughly 700 added cancer deaths in the affected population within 25 years. I do not consider effects other than cancer. The calculation in this second case has the virtue of being related to empirical evidence from fallout which integrates the dose from varied pathways under actual conditions.

Land usage will be compromised by deposited cesium. The deposit of Cs-134 is governed by its radioactive half-life of 2-3 years and that of Cs-137 by an effective mean life in the environment of 14 years on the surface. The external dose, taking into account the shielding effects of buildings and body tissues, associated with radiation from ground-level Cs-137 is roughly 33 mrem per year per Ci/km^2 on the surface,[22] i.e., four Ci/km^2 yields an external dose comparable to background. The impact of deposited cesium on crops is small, but the effect on milk and on meat from grazing animals is extremely large during the first six months or so of deposit. A maximum dose commitment of 8 rem per Ci/km^2 of Cs-137 is indicated for a child via the milk path.[24] About four times as much cesium is concentrated in a kilogram of beef as in a liter of milk.[23]

6.6 Conclusion

Cesium, a semivolatile fission product, has been shown to present a serious hazard during the transportation of spent fuel. The hazard is greater than that due to the noble gases and halogens during transportation. Thus, with rather similar assumptions, a release of 850 curies of Kr-85 has been projected in a loss-of-cooling-water event involving the IF-300 cask.[4] This amount of Kr-85 will have very minor impact compared to the cesium release discussed in the present report. There are a variety of possibilities for corrective action at this time. For example, two aspects of the release of cesium to which the amount of release is sensitive are the use of a fluid under pressure as heat transfer medium and the decay-heat rate, which in turn depends on the cooling-off period for the fuel. A more sweeping modification has been proposed to eliminate all problems associated with long range transport: geographical concentrations of nuclear facilities.[25]

Many of the assumptions made in this chapter are subject to great uncertainty and to variation from case to case. I have attempted to strike a middle ground between underestimating the problems and overestimating them. Some of the numbers estimated here are subject to considerable refinement through detailed calculation. It is hoped that these refinements will be undertaken.

As for many problems associated with the nuclear industry, the one discussed in this chapter could be resolved by appropriate action. It is most disturbing that the cesium problem has not been acted upon by the AEC and in fact has not even been noticed, at least in public print. In my investigation of transportation problems, the escape of radioactive cesium was adopted as the subject for close investigation after only a cursory examination of the possibilities. This reinforces my impression that there are many such problems, and that therefore the nuclear industry is expanding too rapidly. The AEC and nuclear industry have made massive efforts at design and safety. Unfortunately these efforts do not appear to be sufficient when matched against the needs created by the rapid expansion of the industry.

Acknowledgements

I acknowledge with thanks discussion with T. Cochran, D. Geesaman, C. Kikuchi, G. Parker, E. Radford, L. Shappert, and G. Yadigaroglu. These people are, however, not responsible for the manner in which any information was treated. I would also like to thank Joan Anderson and John Davidson for their help and John Gilligan for many essential contributions.

Notes and References

1 A. Reinking, M. Mishouil, G. Yadigaroglu, and V.E. Schrock, "Transport of Nuclear Fuel," Appendix V of "Public Health Risks of Thermal Power Plant" UCLA-ENG-7242, May, 1972, and G. Yadigaroglu et al., "Spent Fuel Transportation Risks," Nuclear News Nov. 1972, p. 71.

2 Environmental Survey of Transportation of Radioactive Materials to and From Nuclear Power Plants, USAEC Directorate of Regulatory Standards, Dec. 1972.

3 Environmental Survey of the Nuclear Fuel Cycle, USAEC Directorate of Licensing, Nov. 1972.

4 "Safety Evaluation by the Transportation Branch Directorate of Licensing," General Electric Co., Model IF-300, Shipping Cask, USAEC 1973.

5 "Siting of Fuel Reprocessing Plants and Waste Management Facilities," ORNL 4451, July 1970.

6 The Safety of Nuclear Power Reactors, USAEC Wash 1250, July 1973.

7 W.A. Brobst, "Transportation Accidents: How Probable?" Nuclear News May 1973, p. 48.

8 L. Shappert, private communication, Nov. 1973, confirms that a loss of the cooling water is to be expected from time to time with many large water-cooled casks.

9 L. Shappert, Cask Designer's Guide, ORNL-

NSIC-68, Feb. 1970.

10 G.W. Parker et al., "Out-of-Pile Studies of Fission Product Release from Overheated Reactor Fuels at ORNL, 1955-1965," ORNL 3981, July 1967.

11 J.L. Bates, "Fission Product Distribution in Irradiated UO_2," Battelle Northwest Laboratory BNWL-58, March 1965.

12 J.H. Davies, R.F. Boyle, and J.F. Hanus, "Fission-Product and Heavy-Element Distributions in High-Performance UO_2 Fuel Rods," ANS Winter Meeting, Nov. 1965.

13 General Electric Co., Reactor Fuels and Reprocessing Dept. Report on IF-300 Shipping Cask, NEDO-10084-1 and Docket 70-1220. I single out this cask not because we know it to present the most serious hazard, but because extensive information is available. A great deal of painstaking design and analysis has gone into the IF-300 cask.

14 C.E. Junge, _Air Chemistry and Radioactivity_, Academic Press, New York, 1963, Chapter 2.

15 J.D.B. Lambert, L.A. Neimark, and R.V. Strain, "A Failure Mechanism in Mixed-Oxide Fuel Elements"; I. Johnson and C.E. Johnson, "Cesium Interaction with Irradiated Oxide Fuel Pins"; and M.G. Adamson and E.A. Aitkin, "Chemical State and Thermomigration Behavior of Fission Product Cesium in Oxide Fuel System," American Nuclear Society Transactions, 1973 Winter Meeting, San Francisco, pp. 193-195.

16 This is the most gusty of class D of M.E. Smith and I.A. Singer, "An Improved Method of Estimating Concentrations and Related Phenomena from a Point Source Emission," J. of Applied Meteorology 1, October, 1966, p. 631.

17 International Commission on Radiological Protection, Report of Committee IV, ICRP Publication 10, Pergamon Press, New York, 1968.

18 This is a weighted dose equivalent for 2/3 Cs-134 and 1/3 Cs-137 as a result of short period inhalation or ingestion event based on the Cs-137 coefficient in the ref. cited in note 17.

19 National Commission on Radiation Protection, Report No. 39, "Basic Radiation Protection Criteria," 1971.

20 Advisory Committee on the Biological Effects of Ionizing Radiations, "The Effects on Populations of Exposure to Low Levels of Ionizing Radiation," National Academy of Sciences, Nov. 1972.

21 In general, particular groups in the population may have a much different susceptibility than the mean of any test population. See, e.g., I. Bross and N. Natarajan, "leukemia from Low-Level Radiation," The New England Journal of Medicine 287 107 (July 20, 1972).

22 United Nations Scientific Committee, "Ionizing Radiation: Levels and Effects," Vol. I, United Nations, New York, 1972.

23 United Nations Scientific Committee,

"Report of the United National Scientific Committee on the Effects of Atomic Radiation," Official Records: Twenty-First Session, Supplement No. 14 (A16314), 1966, pp. 62, 67, 85.

24 Y.C. Ng et al., "Prediction of the Maximum Dosage to Man from the Fallout of Nuclear Devices IV. Handbook for Estimating the Maximum Internal Dose from Radionuclides Released to the Biosphere," UCRL-50163 Part IV, 1968. See also ref. cited in note 22.

25 A. Weinberg, "The Moral Imperatives of Nuclear Energy," Nuclear News Dec. 1971, p. 33, and "Energy Parks: Selection of Site for Study and Proposed Scope of Study," Internal Document, Office of Planning and Analysis, USAEC, Nov. 1972.

Chapter 7

NUCLEAR FUEL REPROCESSING: RADIOLOGICAL IMPACT OF WEST VALLEY PLANT

Philip M. Hatfield

7.1 Summary and Recommendations

In this chapter the operations of the Nuclear Fuel Services (NFS) reprocessing plant, located in West Valley, New York, are described and its radiological impacts on its surroundings for its first five years of operation are reviewed.

The radioactive releases to the air and water from the plant during the period 1966-1971 were significant enough for the AEC and the State of New York to have required major modifications. As a result of the installation of a low-level waste treatment facility, some of the liquid radioactive emissions, since 1971, have been reduced. Nevertheless, the future plant capacity at NFS will eventually increase by a factor of 15 over 1971 throughput. Also, most of the fuel reprocessed at NFS to data has been cooled for much longer than the 150-day minimum required by the NFS license, suggesting that future emissions may be greater as cooling periods are shortened.

The following recommendations are based on the analysis, presented later in the chapter, of the plant's past operations:

1. The more rigorous standards suggested by the National Academy of Sciences-National Research Council be implemented. Specifically, NFS should be permitted to contribute no more than a small fraction of the total permissible population exposure. Ten percent

of the current limit would not be unreasonable.

2. Rather than applying maximum permissible concentrations at the point of discharge into Cattaraugus Creek, the more stringent standards should be applied to Buttermilk Creek, as well as to the on-site lagoons.

3. More rigorous and detailed monitoring of the wildlife within the property boundaries should be undertaken, and the more stringent standards that are applied to humans should also be applied to all major species found within these confines.

4. The emission of I-129 from the plant should be reduced, or better, eliminated entirely. At the very least, the iodine scrubbing system must be operative during all dissolutions. Extensive study of the biological dispersion of I-129 at all trophic levels, both aquatic and terrestrial, is imperative.

5. A krypton recovery system should be added. The emission of long-lived radioactive emitters such as Kr-85, in large quantities, should be reduced or, better, eliminated.

6. Extensive human monitoring should be undertaken to determine the extent of contamination already present, and to detect future increases before unnecessary and excessive exposures occur. The work and dietary habits of the local population should be accurately and systematically determined. A thorough accounting of the I-131, I-129, Pu-238, Sr-90, Cs-137, and Cs-134 content of the tissues of neighboring human populations should be accomplished through post-mortem examination, and correlated with place of residence, work, diet, etc. A program of whole-body counting of selected individuals whose exposure is likely to be excessive should be instituted.

Bioassay programs to estimate tritium, Sr-90, and plutonium contamination of local populations would also be of value.

7. The extent of local groundwater contamination by the lagoons and burial grounds should be determined and monitored.

8. Factors promoting resolubilization and dispersion of radioactive bottom sediments should be assessed, and avoided. The cause for the amount of uranium leaving the lagoons to exceed the amount entering should be investigated and controlled.

7.2 Introduction

Until 1966 all nuclear fuel reprocessing was accomplished at government-owned installations such as the Hanford Works at Richland, Washington, the National Reactor Testing Station in Idaho, and the Savannah River, Georgia, plant. As early as 1960, one private firm - Nuclear Fuels Services, Inc. (NFS) - concluded that recovery and reuse of plutonium and uranium from spent nuclear fuel would be a profitable endeavor.[1]

Based on projected requirements for plutonium in breeder reactors, as well as the recycling of plutonium into currently uranium-fueled light-water reactors, the decision to proceed with the first private reprocessing plant in the world was made in the early sixties. A permit was issued to NFS by the Atomic Energy Commission (AEC) for construction of the West Valley, New York, plant in April 1963, and a provisional operating permit was granted on April 19, 1966. Capital expenditure for this construction was approximately $32 million.[1]

7.3 Plant Site

NFS is located on a 3300 acre site owned by and leased from the New York State Atomic and Space Development Authority near the towns of Ashford and West Valley in Cattaraugus County, New York, 30 miles southeast of Buffalo. Both the property and the reprocessing plant, which lies 4000 feet from the nearest boundary, are isolated from the environs by exclusion fences. The installation sits amidst rolling hills reaching 500 feet above sea level which are cut by two deep ravines to the north of the complex. Erdman's Brook flows by the solid-waste burial grounds southeast of the plant (Figure 7.1) and by the reprocessing plant before emptying into Frank's Creek. The latter drains into Buttermilk Creek, which flows through a 100-foot deep ravine north of the plant. Buttermilk Creek, in turn, flows into Cattaraugus Creek which traverses a 300-foot deep ravine approximately 3 miles north of the plant. The confluence of Buttermilk and Cattaraugus creeks lies just within the limit of the NFS property.

7.4 Population Density

Population density in these environs is relatively low, despite the proximity to Buffalo. Based on 1960 census estimates, there are approximately 6000 people within a 5-mile radius of the plant, 16,000 within 10 miles, and 170,000 within a 24-mile radius.[3] Local wildlife consists of typical upstate New York lowland small game, fowl, and deer. The densities of these species is not known. Aquatic biota in the discharge streams consists of an algal-based food chain supporting a variety

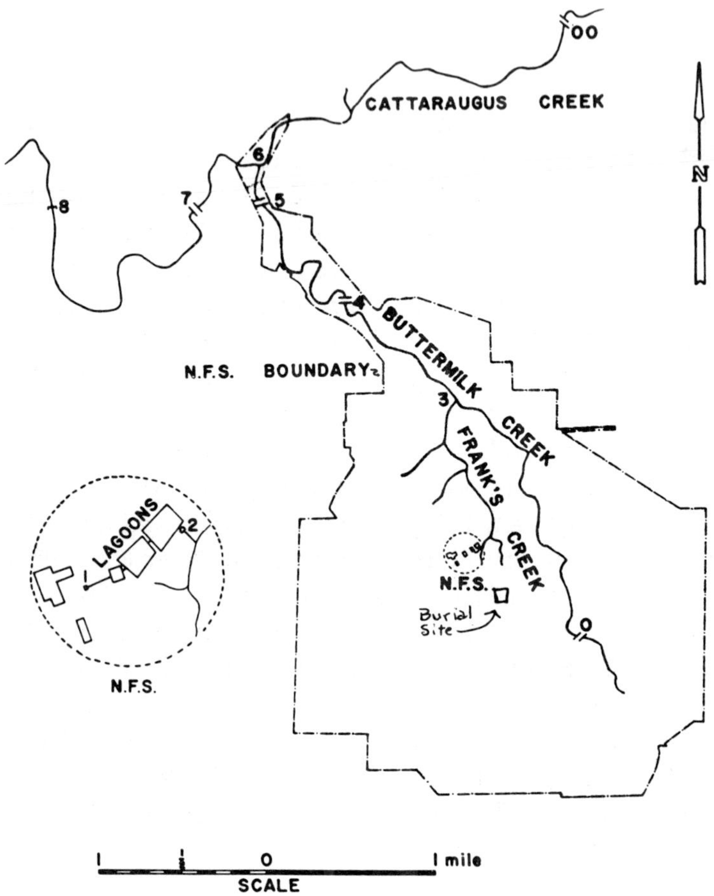

Figure 7.1 Map of NFS fuel reprocessing facility and drainage creeks.

of species of fish, such as shiner, horned dace, chub, white suckers, stoneroller minnows, smallmouth bass, pike, trout, etc. Maximum fishing pressure on the discharge streams occurs at Springville Dam, which lies approximately 6 miles downstream (4 miles as the crow flies) and at the mouth of Cattaraugus Creek on the shores of Lake Erie, approximately 39 miles downstream.[3]

7.5 Uranium and Plutonium Recovery

NFS employs a chop, leach, Purex-solvent extraction process for recovering usable uranium and plutonium from burnt-up reactor fuel. A 1000-megawatt electric (MWe) reactor contains approximately 100 tons of uranium as fuel, one-third of which, or about 30 tons, is replaced annually. Through 1970, NFS employed 190 people who processed over 530 tons of fuel with recovery of 1300 kg of plutonium at a cost to the utilities, in 1969, of $33,380 per metric ton of uranium (MTU).[1] The plant ceased dissolving fuel in November 1971 so that major renovation and expansion could be effected. An average of approximately 100 tons were processed during each of the first 5 years of operation.[1]

By 1975 about 1125 tons of spent fuel should be being processed annually by NFS and any new additional reprocessing plants; and by 1980, 2950 tons.[1] Current expansion plans will lead eventually to a 3 ton per day capacity at NFS.

Spent fuel assemblies are unloaded from their shipping casks underwater, and are placed in a storage pool for further cooling. In the initial processing step, the fuel is transferred to hot cells where the end pieces are placed in baskets. In the subsequent leaching step, the fuel sections are dissolved in a vat of boiling nitric acid which leaves the undissolved fuel containers (cladding hulls) behind. The dissolved fuel is then run through several extraction cycles to separate uranium and plutonium from waste fission products, while the undissolved cladding hulls and other solid wastes are buried in

AEC-licensed pits located just southeast of the processing plant. In the final purification cycle, uranium and plutonium are further refined by additional solvent extraction and ion exchange.[2]

The processing of a single load of fuel from a given reactor is known as a "fuel campaign." In a typical campaign in 1967, 20 tons of fuel from the Yankee plant at Rowe, Massachusetts, were divided into 23 batches for dissolution between the 24th of May and the 9th of July. A given batch is processed, on the average, by 6 hours of dissolution on two consecutive days. The average campaign has required 4-6 weeks to complete. Of the 109 tons processed in 1969, 28 tons were from the New Production Reactor at Hanford and were dissolved between February 17 and April 1; 21 tons from the same source between April 1 and May 10; 20 tons from Rowe Yankee between May 24 and July 9; and an additional 30 tons from Hanford between August 18 and September 21. Relatively low burn-up fuel (on the order of 22,000 megawatt days per metric ton of uranium, or MWD/MTU) was processed from 1966 to 1968.[2] Starting in June 1969, fuels with shorter storage times were processed, resulting in the discharge to the environment of greater amounts of radioactive waste products.[4]

7.6 Radioactive Waste Content of Spent Fuel

Typical nuclear fuels, after 30,000 MWD/MTU burn-up and 150 days' cooling contain 5 million Curies (Ci) of total beta-gamma activity per ton, including 98,000 Ci of strontium-90 (Sr-90), 8400 Ci of krypton-85 (Kr-85), 375 Ci of tritium (H-3), 2.2 Ci of iodine-131 (I-131), and 30 millicuries (30 mCi or 0.03

Ci) of iodine-129 (I-129).[1] Table 7.1 summarizes the total waste inventory of higher burn-up fuel per ton. The fission inventory of spent fuel increases proportionally with fuel burn-up as do the waste discharges. The yields of Kr-85, for example, range from 6000 Ci per ton at 20,000 MWD/MTU burn-up, and 8000 Ci per ton at 35,000 MWD/MTU, to 9100 Ci per ton for 40,000 MWD/MTU.[5] The content of I-129 increases from 30 mCi per ton for 30,000 MWD/MTU to 40 mCi for 40,000 MWD/MTU.[6,7] There are concomitant increases for tritium and other fission products. It is expected that with the increased plant capacity and processing of higher burn-up fuels the tritium released in the NFS liquid discharges will increase from the present high of 1200 Ci per month to as much as 20,000 Ci per month.[1] The uranium contents of spent fuel decreases from the original enriched levels of up to 3% to 0.65-0.85% at the time of reprocessing.

Almost all of the waste Kr-85 is released to the atmosphere during processing along with approximately 25% of the available tritium.[2,8] Smaller amounts of xenon-133 (Xe-133) and Xe-131 and 90% of the I-129 content of the fuel are also discharged through the 60-meter stack.[6] Most of the waste fission and activation products remain in liquid form at the end of reprocessing.

Higher-level liquid wastes are reconcentrated and stored in multiple-walled underground storage tanks set within steel saucers and supplemented by emergency stand-by tanks. About 600,000 gallons of high-level wastes have been accumulated to date. NFS estimates that off-site shipment of these 600,000 gallons, for permanent storage, would cost a prohibitive

Table 7.1 Radionuclide content of light water reactor fuel decayed 150 days.

Nuclide	Concentration (Ci/ metric ton)	Nuclide	Concentration (Ci/ metric ton)
H-3	692	I-131	2.17
Kr-85	11,200	I-132	-
Sr-89	96,000	Xe-133	-
Sr-90	76,600	Cs-134	213,000
Y-90	76,600	Cs-136	20.8
Y-91	159,000	Cs-137	106,000
Zr-95	276,000	Ba-140	430
Nb-95	518,000	La-140	495
Mo-99	-	Ce-141	56,700
Tc-99m	-	Ce-144	770,000
Tc-99	14.2	Pr-143	694
Ru-103	89,100	Nd-147	51.0
Ru-106	410,000	Pm-147	99,400
Rh-103m	89,100	Pm-149	-
Ag-111	-	Sm-151	1150
Cd-115m	44.3	Eu-152	11.5
Sb-124	86.3	Eu-155	6370
Sn-125	20.0	Tb-160	300
Sb-125	8130	Np-239	17.4
Te-125m	3280	Pu-238	2810
Te-127m	6180	Pu-239	330
Te-127	6110	Pu-240	478
Te-129m	6690	Pu-241	115,000
Te-129	4290	Am-241	200
Te-132	-	Cm-242	15,000
I-129	0.038	Cm-244	2490

Source: ORNL-4451

$17 million.[9] Consequently, a high-level liquid waste solidification capacity is included among the current expansion plans. Permanent long-term storage of the solidified high level wastes will be at a federal repository, whenever one is established. In the meantime, these wastes will remain at NFS in either a liquid or solid form. Another section of this book deals further with the waste disposal problem.

7.7 Interceptor Tanks and Lagoons

The low-level liquid wastes from evaporator condensates, plant floor drains, and laundry and decontamination operations are discharged into two 22,000-gallon interceptor tanks.[10] The plant license specifies (spec. 4.2.1 of license CSF-1) that in any calendar quarter no more than 48 Ci of unidentified alpha and beta activity (including tritium and I-131) shall be released from the interceptor tanks, and that the concentration of these wastes shall not exceed 5,000,000 picocuries/liter. (pCi/l; 1 pCi = 10^{-12} Ci, or one-trillionth of a curie.) If the waste concentration in the interceptors exceeds these limits, the wastes are either diluted further or reprocessed prior to discharge to a series of three unlined lagoons excavated from the ground near the interceptor tanks. These lagoons provide a reservoir of approximately 5 million gallons, enough for 100 days' operation.[1] It is estimated that a plant utilizing the Purex process produces between 10,000-100,000 gallons of waste per ton of fuel.[11] Lagoon-1 (300,000-gallon capacity) overflows to lagoon-2 (2,300,000 gallons), which is in turn discharged to lagoon-3 (2,500,000 gallons).[10] The three lagoons, on the average, provide a one month hold-up for the low-level liquid wastes, during which varying amounts of radioactivity settle to the lagoon bottoms. Finally, lagoon-3 is discharged to Erdman's Brook through a monitored valve system at a rate dependent upon the flow in Cattaraugus Creek. Typical discharge rates during the 1969 campaigns were 11.5-75 gallons per minute (up to 108,000 gallons per 24 hours at continuous sustained discharge).[10]

The NFS license stipulates (spec. 4.2.2) that "activity released to Cattaraugus Creek from all sources on the site shall not at any time result in a concentration as measured in the monitoring station in Cattaraugus Creek, which exceeds twice the concentrations specified in Title 10, CFR part 20; provided, however, that in any 365 consecutive days, the average concentration of radioactivity in Cattaraugus Creek shall not exceed that specified in Title 10, CFR part 20." Since both Erdman's Brook and Buttermilk Creek remain within the confines of the NFS restricted area, the AEC sets no limits on the amount of radioactivity in these waters. Concentration limits apply only after dilution in Cattaraugus Creek.

7.8 Liquid Waste Discharges

Table 7.2 summarizes the total tritium, gross alpha, gross beta (less tritium), Sr-90, and I-129 discharged to Cattaraugus Creek in the first 4 years of NFS's operations.[1] The increased quantities in each category are a reflection not so much of increased volumes of fuel processed as of the processing of higher burn-up fuels. Assuming that 100 tons of fuel

Table 7.2 Total annual liquid releases according to figures provided the JCAE by NFS in 1970.

	1966 (Ci)	1967 (Ci)	1968 (Ci)	1969 (Ci)
Gross α	0.038	0.56	0.139	0.376
Gross β	8.265	30.70	46.400	136.100
Tritium	289.520	4204.000	2612.000	5980.000
Sr-90		4.366	5.03	5.289
I-129		0.070	0.028	0.184

contain 37,500 Ci of tritium, and that 65% or approximately 25,000 Ci should be available for discharge to surface waters during reprocessing, the discrepancy between the maximum 5,980 Ci discharged to Cattaraugus Creek in 1969 is attributed to tritium leakage during reactor operations and discharge at the reactor site, before the fuel reaches NFS.

During the February-September campaigns in 1969, a total of 109 tons of fuel were processed, with tritium accounting for 13 times more activity in the lagoons and 33 times more in the Creek than any other single nuclide. The relatively greater tritium concentration in the Creek as compared to the lagoons is accomplished by sedimentation of nuclides such as Sr-90, cesium, Co-60, and Ru-106 in the lagoons.[10]

7.9 Sedimentation In the Lagoons and Creeks

Approximately one-half of the gross beta (excluding tritium) and two-thirds of the alpha activity is retained in the lagoons. Greater than 90% of the Ce-144, Pm-147, and Zr-95 is deposited in the lagoons, as are 70-75% of the cesium isotopes, 59% of the Co-60, 39% of the Ru-106, and 7% of the Sr-90. Among the alpha emitters, sedimentation removes more than 90% of the plutonium discharged from the interceptors. However, U-232, -234, and -238, which account for only 2½% of the activity in the interceptors, are present in 8.5 times greater concentration at the point of discharge from lagoon-3, indicating that uranium previously deposited in the lagoons is redissolved and discharged to Cattaraugus Creek. Most of the Ru-106, Sr-90, Co-60, and uranium are present in a soluble form in the lagoons, accounting for their relatively greater concentration at the point of discharge into Erdman's Brook. Those nuclides which are deposited in the lagoons are largely in an insoluble form, except the cesium isotopes, which may be either soluble or insoluble.[10]

Tritium, Ru-106, and Sr-90 constituted the principal nuclides in both Buttermilk and Cattaraugus Creeks during these campaigns. Lesser amounts of Cs-134, Co-60, U-232, -234, and -238 are also found in the Creeks. During these 7 months of 1969, a total of 1700 Ci of tritium, 15 Ci of Ru-106, 8.3 Ci of Sr-90, 8 Ci of Cs-137, 2 Ci of Cs-134, 0.16 Ci of Ce-134, 0.2 Ci of Co-60, 0.59 Ci of Sb-125, etc. were discharged from lagoon-3 to Erdman's Brook. Among the alpha emitters, a total of 1.8 mCi of plutonium and 32 mCi of uranium were released.[10]

Greater than 90% of the Cs-137 activity was removed from the water by deposition in passage from lagoon-3 to Cattaraugus Creek. Erdman's Brook and Frank's Creek also retained 66% of the Co-60, 72% of the Sr-90, 31% of the Sb-125, and 26% of the Ru-106. Deposition within the property boundaries, of the Sr-90 at least, was quite variable and pH dependent. In November, when the waste streams were neutral, only 10% of the Sr-90 was removed, whereas in June 75% of the Sr-90 was deposited when the waste-stream[10] pH was 11. Other parameters affecting deposition were not investigated, nor is much known about what factors promote dissolution of elements previously deposited in the lagoons.

7.10 Monitoring by the State of New York

The New York State Department of Environmental Conservation instituted a radiation surveillance program around NFS soon after startup. The New York State data are usually expressed as average, minimum, and maximum concentrations measured at several points along Buttermilk and Cattaraugus Creeks.[4,12-16] Tables 7.3 and 7.4 summarize the maximum values observed in Buttermilk Creek at Thomas Corner Road and in Cattaraugus Creek at Springville Dam (6 miles from lagoon-3 and 2 miles from the confluence of Buttermilk and Cattaraugus Creeks) from 1966 to 1972. Maximum values are recorded in these tables because radiation protection standards are designed to provide a margin of safety above and beyond the highest observed levels. The 1966 Cs-137 level represents the activity of this isotope before the plant became operational.

Table 7.3 Maximum isotopic levels in Buttermilk Creek 1966–1972 in pCi/l (New York State data).

	Tritium	Sr-90	Cs-137	Ru-106	I-129	Pu-238	Pu-239
1966			1				
1967			100				
1968	1,133,920	1,015	666				
1969	431,010	1,658	1,161	5,816		0.63	0.60
1970	1,185,290	2,008	972	3,917		2.40	1.90
1971	1,480,630	1,841	692	10,875	7.8	1.80	0.078
1972	90,200	357	41	3,846	1.4	0.23	0.08

Table 7.4 Maximum isotopic levels in Cattaraugus Creek at Springville Dam 1966–1972 in pCi/l (New York State data).

	Tritium	Sr-90	Cs-137	Ru-106	I-129	Pu-238	Pu-239
1966			1				
1967			10				
1968	110,430	60	20				
1969	98,600	224	86	421		0.64	1.36
1970	103,770	291	197	782		3.60	2.20
1971	184,510	117	29	481	0.6	0.016	0.014
1972	9,090		119				

7.11 Maximum Permissible Water Concentrations

The maximum permissible concentrations allowed in plant effluent water (MPC_W) by the State of New York are indicated in Table 7.5.[12]

Table 7.5 State of New York maximum permissible concentrations in plant effluent water (mpc_w).[12]

Isotope	MPC_W (pCi/l)
Tritium	3,000,000
Sr-90	300
Cs-137	20,000
Cs-134	9,000
Co-60	50,000
Ru-106	10,000
I-129	60
Pu-238	5,000
Pu-239	5,000
U-234	30,000
U-235	30,000

These concentrations correspond to maximum permissible levels established by the federal government in 10 CFR part 20. Ingestion of a given amount of water or food per year, containing these maximum concentrations of each isotope, if it alone is present, would result in the maximum permissible radiation dose of 500 millirems per year to the whole body (by tritium and cesium), 500 millirems per year to bone (by Sr-90, Pu-238, and Pu-239), and 1500 millirems per year to the stomach and intestines (by Co-60 and Ru-106).[17] The maximum permissible thyroid dose of 1500 millirems per year would be achieved by drinking average amounts of water containing 60 pCi of I-129 per liter every day of the year. It should be noted that all such limits are based on standard adult parameters, and assume that only a single isotope is present. If, for example, the Cs-137 level in water results in a whole body exposure of 125 millirems per year, and the Cs-134 another 125 millirems, then the maximum permissible level of tritium would have to be reduced by one-half so that the total would still be 500 millirems. Children tend to accumulate some of the isotopes more avidly than adults. The whole body dose to an infant ingesting Cs-137 is almost twice that to an adult consuming the same amount of cesium.[18]

7.12 Radioactive Contamination of Creeks

It is clear that from start-up in 1966 through 1970 there was an increase in the levels of all the isotopes in both Buttermilk and Cattaraugus Creeks. It should be emphasized that although the maximum levels of SR-90 have often exceeded the average permissible concentration (300 pCi/l) in Buttermilk Creek, in the view of the AEC the limits were not exceeded since Buttermilk Creek is within the NFS restricted area which is exempted from the 10 CFR part 20 regulations. The standards are not applied until Buttermilk Creek empties into Cattaraugus Creek just within the NFS boundary. Unfortunately, the wildlife in the area are unaware of this regulation.

It has been said by those responsible for the plant that the levels in Cattaraugus Creek "have not exceeded 20% of the maximum allowable concentrations."[1,9] What is not said is that the statement is true only when the average annual concentration for single nuclides are compared with the average annual limits. Concentrations measured over shorter periods, however, were found to be quite substantial. From Table 7.4, for example, we see that on one occasion in 1970 the Sr-90 level in Cattaraugus Creek reached 97% of the maximum permissible limit. Further, on NFS property - where no limits are imposed - the maximum Sr-90 values in Buttermilk Creek would have consistently exceeded the limits from 1968 until 1971 (see Table 7.3). Iodine-129 and plutonium are also present in the creek, and increased markedly from 1969 to 1970. The State of New York noted a significant rise in Sr-90 concentrations in both creeks in the last 3 months of 1969.[4] As stated in the 1970 annual report from the Department of Environmental Conservation, "the amounts of radioactivity detected in water and fish remained at levels which confirmed the need to install liquid waste treatment facilities to hold discharges as low as practicable."[13] This comment followed an admonition, in the 1969 report, that the "concentration of radioactivity ...particularly Sr-90, from the Creek systems below the discharge at the fuels reprocessing plant, remain at levels that indicate that corrective action should be taken in order to hold liquid waste discharges to as low levels as practicable."[4] Corrective action was finally initiated by the AEC at the request of the State of New York. A new low level waste treatment plant became operative on May 23, 1971. Additional filtration, sedimentation, and ion exchange processes, with burial of the contaminated waste sludge and spent resins in the on-site burial ground, resulted in the lower levels of Sr-90 and cesium relative to earlier years. The increase in tritium, which is not removed by any existing waste treatment process, indicates that without the new facilities, the levels of Sr-90 and cesium would also have increased in 1971. Ru-106 levels were not particularly reduced. Consistently low levels were observed in 1972 since no fuel was processed during the year.

Prior to 1952, the deposition of naturally occurring tritium in the atmosphere produced by cosmic ray interactions produced a natural tritium concentration of 6-35 pCi/l in surface waters. Atmospheric and, consequentially, surface water concentrations increased markedly during the period of active nuclear weapons testing. In order to monitor these levels, the Bureau of Radiologic Health has maintained ten stations around the United States which have routinely analyzed tritium contamination for many years. From 1964 to 1966 these levels ranged from 2000 to 20,000 pCi/l.[19] The maximum level of 1,480,000 in Buttermilk Creek in 1969 is far higher than values observed elsewhere in the United States, including rivers downstream from nuclear installations. The high "normal" value of 20,000 pCi/l in the surface waters of the United States corresponds to a whole body dose of 1.14 millirems per year assuming ingestion of 440 liters of water per year. Although humans are not known to use Cattaraugus Creek for drinking water, an animal drinking Cattaraugus Creek water at the maximum observed tritium level (184,000 pCi/l at Springville Dam) would receive 10 millirems whole body irradiation per year. The average tritium content of Cattaraugus Creek at Springville Dam increased from

19,328 pCi/l in 1970 to 30,994 in 1971, and may be anticipated to increase steadily in the future. While the tritium contamination of drinking water is not routinely monitored throughout the State of New York, several community water supplies around NFS have been analyzed in recent years. There has been no detectable tritium in this water to date (minimal detectable concentration is 500 pCi/l). No monitoring wells exist around the plant to determine the extent of contamination of groundwater, if any.[1] A surface monitoring site, upstream from NFS, has revealed seepage from the burial grounds, however. Tritium concentrations here reached a maximum level of 426,490 pCi/l in 1971.[14]

Considering the large quantities of tritium released, we believe that it will be only a matter of time until detectable contamination of groundwater occurs. It is imperative that the likelihood and extent of such contamination be determined. Such information could modify current tritium release procedures if unsuspected hazards are found. While NFS is quite aware of these problems, it appears that it is not technically possible to reduce the tritium output at present. Since the amounts of tritium to be released in the future will increase markedly, it is imperative that methods be developed to prevent this contamination.

Maximum plutonium contamination in the creeks in 1971 is indicated in Table 7.6.[14]

7.13 Radioactive Contamination of Fish

Tables 7.7 and 7.8 summarize the maximum levels of Sr-90, Cs-134, Co-60, and Ru-106 in fish taken from Buttermilk and Cattaraugus

Table 7.6 Maximum plutonium contamination, 1971.[14]

	Pu-238 (pCi/l)	Pu-239 (pCi/l)
Buttermilk Creek	1.80	0.78
Cattaraugus Creek	0.016	0.014

Creeks between 1964 and 1971.[4,12-16] Prior to operation of NFS, the majority of nuclides were undetectable. Prior to 1970, whole fish were analyzed (i.e., fish bone and flesh were not separated). Most of the Sr-90 concentrates in bone so that the pre-1970 values are not an accurate estimate of the Sr-90 concentrations in the eatable portions of the fish. The maximum Sr-90 level of 1400 pCi/kg in 1969 compared with the 1970 maximum of 1679 pCi/kg represents a much greater increase than is apparent from the numbers alone. The bone level of Sr-90 in the maximum 1970 sample was 10,491 pCi/kg which, compared to the 1969 level of 1400 pCi/kg, indicates a marked increase.

A general increase in all the nuclides in fish occurred between 1966 and 1970. Bottom feeders such as shiner, dart, dace, and sucker generally manifest higher levels than do game fish such as trout, bass, and pike, which feed nearer the surface. Unlike Sr-90, cesium, cobalt, and ruthenium concentrate largely in the fleshy parts so that pre-1970 whole fish levels are not likely to differ significantly from the 1970 and 1971 values.

It is not difficult to estimate the amount of radiation one would receive if he consumed

Table 7.7 Maximum isotopic levels in Buttermilk Creek bottom feeders 1965-1971 in pCi/kg (New York State data).

	Sr-90	Cs-137	Co-60	Cs-134	Ru-106
1965		1,685			
1966	81				
1967	15,300	27,500	340	5,750	1,750
1968	22,800	13,400	342	3,140	2,990
1969	5,668	6,032	242	689	3,611
1970	14,456	34,835		5,890	
1971		4,659		742	

Table 7.8 Maximum isotopic levels in Cattaraugus Creek at Springville Dam in bottom feeders 1965-1971 in pCi/kg (New York State data).

	Sr-90	Cs-137	Co-60	Cs-134	Ru-106
1965		23			
1966	1,035				
1967	906	3,610	32	499	796
1968	3,080	1,330	382	1,170	155
1969	1,400	3,141	95	1,547	389
1970	1,679	3,198		534	
1971	45	1,666	23	301	

water, fish, deer, or other foodstuffs containing radioisotopes. In the case of Sr-90, the isotope would concentrate mainly in human bone, and the significant radiation dose would be to bone. Cesium would distribute primarily in human muscle, and the dose would be to the whole body. Ruthenium and cobalt would not be well absorbed from the stomach or intestines, so that the dose would be to the lining of the gut. Tritium distributes with the body water and produces whole body exposure.

Since most of the isotopes have long physical half lives, the individual ingesting the material would not receive the entire radiation dose immediately. The cesium isotopes, Ru-106 and Co-60, remain in the body a relatively short time (biological half lives of less than 100 days), so that almost all of the radiation is delivered within the first year after ingestion. Sr-90, on the other hand, becomes more or less permanently bound to bone, so that only 7% of the dose from a given ingestion is delivered during the first year. One must therefore express the radiation dose as the 50-year dose commitment for Sr-90. In the case of the non-strontium isotopes discussed previously, almost all of the radiation is received within the first year so that the dose commitment may be thought of as the annual dose.

If an individual ingests radioactivity on a continuous basis, as to those around NFS, the annual dose can be calculated as a series of individual ingestions. If, for example, a person drinks water, every day, containing a certain amount of Sr-90 (in pCi/l), one calculates the total pCi ingested per year, year by year, by multiplying the water concentration (in pCi/l of water) times the number of liters consumed over the entire year (440 liters on the average). By multiplying this value by the appropriate conversion factor in Table 7.9, the 50-year dose commitment to bone from Sr-90 for that year or that ingestion can be calculated (in millirems per 50 years). If the concentration in meat or fish is known (in pCi/kg), the 50-year dose commitment for any quantity of the material is also readily obtainable, as will now be shown.

The Bureau of Radiological Health estimates that a "typical" individual in the vicinity of NFS consumed 5.5 kg (1 kg = 2.2 lb.) of local fish and 2 kg of local deer meat per year, while the hypothetical "maximum" individual consumes 50 kg of local deer meat and 50 kg of local fish per year.[3] If, for example, one ate 1 kg of fish containing 5400 pCi Sr-90 per kg of flesh (the maximum value in Buttermilk Creek trout in 1970), the total 50-year dose commitment to bone from that one meal would be $5400 \times 8.4 \times 10^{-3} = 45.4$ mrem. Approximately 7% of this dose, or 3.2 mrem, would be received in the first year. (We stress again, however, that at this time fishing is not permitted on the NFS site.) If, over the course of the year, such an individual consumed 10 kg of such fish, the dose commitment would be 32 mrem for that year.

Since measurements of concentrations in fish are difficult and time consuming, it is useful to estimate what concentration a fish living in a given body of water is likely to have. In other words, by calculating the maximum amount that a given fish species has been observed to concentrate an isotope from the surrounding water, it becomes possible to

Table 7.9 Target organ and dose per pCi of each isotope consumed (multiply concentration in media consumed in pCi/l or pCi/kg by column 4 to obtain 50-year dose commitments).

Isotope	Target organ	Method of consumption	Millirems/50 years per pCi of intake (concentration in pCi/l of water, food, or air)
Sr-90	bone	ingestion	8.4×10^{-3}
Pu-239	bone	ingestion	1.2×10^{-3}
Cs-137	whole body	ingestion	0.06×10^{-3}
Cs-134	whole body	ingestion	0.08×10^{-3}
Ru-106	GI tract	ingestion	0.195×10^{-3}
Co-60	GI tract	ingestion	0.04×10^{-3}
Tritium	whole body	inhalation; skin absorption from air	1.86
Kr-85	whole body	inhalation; skin absorption from air	2.08

Table 7.9 Target organ and dose per pCi of each isotope consumed (multiply concentration in media consumed in pCi/l or pCi/kg by column 4 to obtain 50-year dose commitments)(continued).

Isotope	Target Organ	Method of consumption	Millirems/50 years per pCi of intake (concentration in pCi/l of water, food, or air)
Pu-238	Hilar lymph nodes	inhalation	132
Pu-239	Hilar lymph nodes	inhalation	260

estimate the dose received by eating fish taken from that water, knowing only the concentration of the isotope in the water and the amount of fish consumed. It is apparent from Tables 7.3 and 7.4 that the concentrations of the various isotopes in Cattaraugus Creek are considerably less than in Buttermilk Creek, differing, on the average, by a factor of ten (except for plutonium). However, the concentration in fish flesh, relative to the concentration in the surrounding water (the so-called concentration factor) does not differ markedly between the two creeks.

According to measurements made by the State of New York, the concentration of Sr-90 in the flesh of bottom feeders in Cattaraugus Creek may exceed the concentration in the surrounding water by 22 times.[4] Maximum concentration factors have also been determined for Cs-137, Cs-134, Ru-106, and Co-60 by dividing the maximum fish levels by the maximum water concentrations for that year. Concentration factors derived by this method are quite conservative compared to values reported in the literature. Concentration factors as high as 22,000 have been reported for cesium in freshwater fish, 90 for Sr-90, and 3450 for Co-60.[20] Tritium is not concentrated by fish, but occurs at approximately the same level in fish as in the surrounding water. The estimated 50-year dose commitment after consumption of a given number of pounds of fish, knowing only the isotope concentration in the water, can be obtained simply by reference to Table 7.10.

Estimated radiation doses to humans could be much higher if the concentration factors used here, derived from relatively few fish samples, are lower than the actual maximum

Table 7.10 Estimated radiation dosage from consuming x pounds of fish flesh from Buttermilk or Cattaraugus Creeks knowing only the water concentration of the isotope of interest.

		Maximum Concentration factor in		Multiply water concentration in pCi/l by this conversion factor x number of pounds of fish consumed to obtain 50-year dose commitment for	
Isotope	Target organ	Bottom feeders	Surface feeders	Bottom feeders	Surface feeders
Sr-90	bone	22	4.6	0.08	0.017
Cs-137	whole body	54	13.2	0.0015	0.0036
Cs-134	"	25.3	2.7	0.00092	0.000098
Ru-106	GI tract	3.0	25.1	0.00027	0.0023
Co-60	"	95.5	161.5	0.0017	0.0029

values as suggested by the literature. With 2009 pCi of Sr-90 per liter of Buttermilk Creek water (1970) and a maximum concentrating factor of 22, the 50-year dose commitment to the bone of a person eating 10 kg (22 lbs.) of fish from these waters would be 2008 x 10 x 22 x 8.4 x 10^{-3} = 3710 millirems, or 3710 x 7% = 260 millirems per year.

Granted that a worst possible case illustration has been used, it is based on observed concentrating abilities of local fish and observed isotope levels in the discharge creeks. Without firm evidence that edible fish do not migrate from Buttermilk Creek, and realizing that actual maximum concentrating factors may be higher than those given here, it is not unreasonable to ask why standards have not been applied to Buttermilk Creek. The highest recorded value of Sr-90 in the flesh of bottom feeders taken from Buttermilk Creek is 14,456 pCi/kg. Ingestion of 1 kg of such fish would result in a 50-year dose commitment to bone of 121.4 millirems.

7.14 Need for More Rigorous Standards

Although these may be extreme instances, the paucity of information on the actual quantity of fish consumed in the region, the migration patterns of fish, and their isotopic concentration make more restrictive standards mandatory. While the new low-level waste treatment facility has resulted in lower levels in the creeks and fish in 1971, the fact that the plant will be processing higher burn-up fuel in greater quantities in the future may cause a reversal in this downtrend. It is imperative that more restrictive controls be applied, and that the frequency with which these worst possible cases are actually occurring be accurately determined. Bland reassurances that "our record of performance relative to protection of the public and of our own employees has been successful when measured against permissible standards as authorized in our AEC facility license," and that "on an annual basis, our liquid effluents have ranged from 6-18% of Federal regulations with a typical value of 10%,"[1] are numerical half-truths obscuring the degree of contamination that has occurred. What is important is not only what comes out, or how much relative to such and such a standard, but where it goes and what it does once it gets out. Without independent monitoring by the State of New York, the magnitude of the problem would still be obfuscated. As indicated later in the testimony of Mr. Runion (President of NFS) before the Joint Congressional Committee on Atomic Energy in 1970, NFS has not been as successful in controlling emissions as originally anticipated. "We try to maintain as a target in our original design something on the order of less than 5% of the limit. The operating characteristics of the plant, the practical problems that we ran into, have all pushed us above these limits. Fortunately, the plant has not yet operated at a level above 50% of its capacity."[1] And yet, NFS is undergoing a major expansion.

If, as recently recommended by the National Academy of Sciences-National Research Council,[21] the nuclear industry were apportioned only a fraction (say 1/20 to 1/10) of the present maximum permissible population exposures, with the remainder divided up among other sources of radiation exposure, then NFS would be required to limit its emissions to a small percentage of the 1966-1971 levels. If, for example, the plant could not maintain an average Sr-90 level of less than 15-30 pCi/l in Buttermilk Creek, it would be in excess of the limit. It is interesting to note that the more modern Barnwell (South Carolina) reprocessing plant has been designed to emit no liquid effluents. It is not unreasonable to ask why such performance should not be required of NFS.

7.15 Leakage From Burial Grounds

The low cesium concentrations in the creeks in 1966 prior to start-up reflect overflow or seepage from the solid waste burial trenches southeast of the plant.[22] These AEC-licensed pits are excavated from the ground and filled with undissolved fuel cladding and other low specific activity solid wastes. The solid detritus from 1970 operations filled a single 300,000 cubic foot trench (trench #8) with 150,000 cubic feet of waste containing 40,000 Ci of total beta-gamma activity.[9] The trenches are not covered until full, so that the radioactive materials are often soaked by snow and rain. During 1970 approximately 1,000,000 gallons of water were pumped from trench #9. NFS estimates that this water contained only 40 mCi of activity, i.e., only 1/1,000,000 of the total activity in the trench.[9] However, Erdman's Brook, upstream from the processing plant and downstream from the burial site, contains significant amounts of activity leached from the burial site, which, in turn, contributed to the isotopic inventory of Buttermilk Creek. In recent years, a monitoring station on a small brook northeast of the burial site has also demonstrated some activity, indicating either surface drainage or underground seepage in this direction.[14] The isotopic inventory of the burial site is not publicly available, if it is known, and the rate of migration from the burial site has not been determined. There are no monitoring wells in the vicinity, and the extent of groundwater contamination, if any, and the likelihood of such contamination, are unknown.

7.16 Radioactive Contamination of Deer

Contamination of local deer is attributed primarily to deer drinking from the storage lagoons and Buttermilk Creek. Maximum levels of strontium and cesium in deer meat in 1970 and 1971 declined compared to the first three years of operation (see Table 7.11). Cobalt levels remained relatively constant, while tritium content increased. The 50-year dose commitment to persons consuming deer meat containing the maximum levels in 1971 are relatively low. Such meat would produce a 50-year dose commitment to bone of 0.04 mrem per pound of meat, and 0.05 mrem whole body from the Cs-137 and Cs-134. The level of Sr-90 levels present in 1968 would have resulted in a 50-year bone dose commitment of up to 2.2 mrem per pound. Whole body irradiation from the maximum tritium content in 1969 would have been 0.09 mrem per pound to the person consuming such meat.

Although no standards exist in this country protecting the wildlife around nuclear installations, observed levels in deer, were they observed in man, would be cause for concern. Assuming a deer weighs 220 pounds (100 kg), and that the tritium content of the

Table 7.11 Maximum isotopic levels in NFS deer, 1967-1971 (New York State data).

	Sr-90 (pCi/kg)	Cs-137 (pCi/kg)	Cs-134 (pCi/kg)	Co-60 (pCi/kg)	Tritium (pCi/kg)	Thyroid I-129 (pCi/g)
1967		66,956	12,313	151	46,300	
1968	543	3,793	592	15		
1969	95	8,587	2,046	15		1,200
1970	28	4,687	784	56		3,700
1971	10	1,428	314	15	683,550	1,400

entire body in 1971 was the same as that of the meat, the deer could contain up to 683,550 x 100 pCi of tritium. The whole body irradiation to such a deer would be 8.8 mrem. Assuming 30 kg of muscle, the total 1971 cesium (137) content would be 32,840 pCi and the whole body dose 1.97 mrem. At the maximum 1967 Cs-137 level of 66,956 pCi/kg, the whole body dose would have been 120 mrem. It is interesting that the lower levels of cesium in 1971, relative to 1969 and 1970, are still higher than the levels in Colorado mule deer at the peak of atmospheric fallout in 1963.[23] Alarm over the ubiquitousness of this radioactive contamination of all life forms set the stage for the successful negotiation of the nuclear weapons test ban treaty in 1963. Since that time, Cs-137 levels in Colorado deer have declined to an average of 64 pCi/kg, or twenty times less than in deer around NFS in 1971.[23]

7.17 Iodine-129 Contamination

The level of I-129 in Buttermilk Creek reached a peak of 7.8 pCi/l in 1971 and declined to 1.4 pCi/l in the first six months of 1972.[14-16] Approximately 2% of the I-129 content of spent fuel is discharged to the liquid effluent, and the balance to the atmosphere.[6] The great majority of the vaporized I-129 can be retained on iodine absorbers. However, the plant is permitted to bypass these absorbers if the I-131 content of the fuel is low. Since much of the fuel has been stored for long periods, the I-131 content has been sufficiently low so that the scrubbers have not always been operative during dissolution.[14] The result has been unnecessary discharge of I-129 to the atmosphere. With a half life of 17 million years, I-129 release constitutes an irretrievable environmental degradation. Most of the I-129 released to the atmosphere is returned to the ground within 20 miles of the plant.[6] The fate of this material once deposited on the ground is largely unknown.

According to the specific-activity method, a thyroid gland containing 0.03 μCi (1 μCi = 10^{-6} Ci) of I-129 will receive the maximum permissible annual thyroid dose of 1.5 rem.[6] Assuming a 30-gm thyroid in deer, the 1970 maximum thyroid I-129 concentration of 3700 pCi/gm would deliver 5.6 rem, or almost 4 times the human limit, to the deer thyroid. Fortunately, deer thyroid is not known to be consumed by most hunters. It is unclear precisely how these deer obtain such quantities of I-129. If a deer took all of its drinking water from Buttermilk Creek with a maximum I-129 concentration of 7.8 pCi/l, the total thyroid I-129 content would only have been on the order of 25 pCi/gm, substantially less than the observed maximum. It is unknown what other sources of I-129 are available to deer and small game in the NFS area. One likely possibility is that I-129 is more extensively distributed in the local food chain than is realized. A single specimen of weed taken from Cattaraugus Creek in 1967 contained 0.81 μCi of I-129 per gram of total iodine, presumably derived from the iodine in the creek system.[6] It is not unreasonable to assume that certain biota in the woods around NFS are retaining and concentrating I-129 from the atmosphere in a similar fashion. Further study is necessary to determine the extent of

Nothing is known about the human intake of I-129 or the concentration in thyroid glands of humans in this area. While the concentration in Cattaraugus Creek in 1971 was approximately 0.6 pCi/l, concentrations of I-129 in the local water supplies have not been determined. A human drinking 2.2 liters of Cattaraugus Creek water per day would accumulate approximately 0.00016 μCi of I-129 after one year, and receive a thyroid dose of 5 mrem per year.

The State of New York has maintained a milk surveillance network for several years, but has had the ability to detect I-129 levels of less than 16 pCi/l for only the past two years. In 1971, the last year for which data are available, the maximum milk I-129 concentration was 1.2 pCi/l at a farm 1.7 miles east of the plant.[14] Were 1 liter of such milk consumed every day, the annual human thyroid dose would have been 2.7 mrem per year.

7.18 Atmospheric Kr-85 Contamination

In terms of the total curies of activity discharged through the plant stack, Kr-85 is the most important nuclide by far. During the 1969 fuel campaigns, an average of approximately 5000 Ci of Kr-85 were released per day of dissolution, compared with 40-50 Ci of tritium per day.[2] Twenty to fifty Ci of Kr-85 are released during the initial chopping operation, and the remainder during the dissolution step. Kr-85 discharges from the stack reach a maximum around 10 minutes after the start of dissolution, then decrease toward background over the next 3-4 hours. During 2 days of processing in 1969 (June 12 and 14), the peak atmospheric Kr-85 at the property boundary was 9.3×10^{-6} μCi/cc (or 9300 pCi/l). The average ranged from 4.8×10^{-7} to 6.4×10^{-8} μCi/cc. The average fence limit is 1×10^{-7} μCi/cc, or 100 pCi/l. Actual measurements indicate that by the time Kr-85 reaches the fence, 1500 meters from the stack, it has been diluted by about 42,000 times.[2]

Were the peak of 9300 pCi/l (achieved during dissolution) to prevail at the boundary throughout the year, the whole body dose in the path of the plume would be 19.3 mrem. It is estimated that a 3-hour maximum of this order of magnitude corresponds to a 24-hour average concentration at the fence of 680 pCi/l,[2] which would produce a whole body dose of 1.46 mrem per year if dissolution were undertaken at the plant every day. Table 7.12 summarizes the maximum Kr-85 levels at various monitoring sites maintained by the State of New York in Ashford. Krypton-85 levels in Albany, several hundred miles to the east, ranged from 0.035 to 0.088 pCi/l. The maximum value at Ashford in 1969 was 7.8 pCi/l,[4] which would result in an annual whole body dose of 16.2 mrem if sustained for the entire year. Continuous observations have not been made and the actual doses delivered to the surrounding population are unknown.

Since a variety of techniques are available for removing Kr-85 from the gaseous effluent, NFS commissioned the Blaw-Knox Company, in 1971, to determine the cost of an improved krypton recovery system.[1] Whether the decision has been made to spend the money that such a system would require is unknown. Considering the quantities of higher burn-up fuel that will be processed in the future, it is anticipated that exposures to surrounding

Table 7.12 Kr-85 levels in air in pCi/m^3 1969-1971 at Ashford and Albany (New York State data).

			Average	Maximum
1969	Ashford			
		Site AO54145		7,860
		Site AO45114		2,640
	Albany		21	88
1970	Ashford	Site AO45145		1,640
		Site AO45144		221
	Albany		19	37
1971	Ashford	Site AO45143	1,425[a]	
		Site AO45114	137	1,516
	Albany		18	35

a. Based on only one measurement.

populations will increase significantly if such a system is not installed. It has been estimated that a plant processing 6 MTU of 40,000 MWD/MTU per day of fuel would release 1.2 Ci of Kr-85/sec,[5] compared to an average Kr-85 release rate of 0.047 Ci/sec currently at NFS. If NFS operates at its new capacity of three tons per day when it reopens, Kr-85 releases could increase ten times over current levels, and maximum whole body exposures in Ashford could be as high as 163 mrem per year.

7.19 Atmospheric Tritium Contamination

While the amount of tritium discharged to the atmosphere is small, compared to the quantity of Kr-85, the radiation dose from the tritium is not altogether inconsequential. Since most of the tritium releases from nuclear facilities are into the liquid waste streams, there has been relatively little attention paid to the atmospheric releases. However, it is known that, during the 1969 fuel campaigns, 0.6 Ci of tritium emanated from the stack during the processing of batch #13 on June 12 and 0.4 Ci from batch #14 on June 14.[2]

The residual atmospheric tritium from nuclear weapons testing, together with naturally occurring tritium in the stratosphere, accounts for the ambient tritium levels of 200-1000 pCi/l of moisture in air.[8] The concentration of tritium per liter of air is 1.18×10^{-5} times the concentration of tritium per liter of moisture in air.[3]

Stratospheric tritium remains in the atmosphere for long periods, with a mean residence time of 1 to 10 years, then gradually percolates to lower tropospheric levels, where it is rapidly flushed out by rainfall after an average residence time of one month. For local releases, such as from the NFS stack, the downward diffusion of the plume probably results in the tritium depositing on the ground before washout by rain can occur. Approximately 72% of the tritium on the ground is returned to the atmosphere by evaporation and transpiration every year, with the balance migrating by surface runoff and underground seepage.[24]

In a 1967 study at NFS, over 700 air samples were collected over 7 legs radiating from the stack out to a distance of 4 miles.[8] An arbitrary value of 2500 pCi of tritium per liter of moisture in air was chosen as the upper limit of ambient tritium. Seventy-five percent of the observed values on the leg extending northwest from the stack exceeded 2500 pCi/l, as did 20% on the northern and southern radials. The highest recorded value was 26,400 pCi/l. The tritium concentration exceeded 10,000 pCi/l 39% of the time northwest of the plant. Constant immersion, for 365 days, in air containing 2500 pCi of tritium per liter of moisture in air would produce a dose of less than 0.1 mrem/yr. If we assume an additional 10,000 pCi/l of tritium present from 3 hours of fuel dissolution for 250 days per year then an added whole body dose of 0.03 mrem would be received. Thus the total tritium exposure from air would be 0.13 mrem/yr, small, at present, compared with other exposures from the plant.

Springville Dam lies along the northwest radial, approximately 4 miles from the stack, and fishermen at the Dam would receive a small whole body exposure from the tritium in the air, in addition to the atmospheric Kr-85 contribution and the radiation emanating from the radioactive sediments in the Creek bed. The maximum background exposure rate along the bank of Cattaraugus Creek, in November 1970, was 0.132 mrem per hour.[13]

The authors of the 1967 tritium study concluded that "it is recognized that much more work needs to be done."[8] Such work remains to be done. Two years later, Mr. Runion, the President of NFS, testified before the Joint Committee on Atomic Energy that "airborne contamination has not been a problem at West Valley."[1]

7.20 Atmospheric Plutonium Contamination

In addition to significant quantities of krypton and tritium in the air, there are also measurable quantities of the isotopes of plutonium. This substance has been characterized as "among the most hazardous materials known to man."[25] The maximum permissible air concentration (MPC_a) for Pu-238, of 0.07 pCi/m^3 would result in an exposure to the lungs of 1.5 rem per year. The MPC_a for Pu-239 is 0.06 pCi/m^3. Once a person breathes plutonium-laden air, a certain portion of the plutonium, determined by the particle sizes, is trapped in the lungs with a biological half life of 500 days. After a certain time, most of the lung plutonium is cleared to the lymph nodes about the lung roots where it remains more or less permanently. The maximum lung radiation is to these so-called hilar lymph nodes.

A small quantity of plutonium persists in the air from nuclear weapons tests, and is characterized by a Pu-239: Pu-238 ratio of much greater than 1, commonly from 5 to 15.

Plutonium around the West Valley plant, to be definitely attributed to plant operations, must manifest a Pu-239:Pu-238 ratio of less than 3, or must manifest much higher levels than might be expected from fallout alone.[13] Low burn-up fuel, such as processed in the earlier campaigns, is characterized by a ratio of from 1 to 3, whereas the higher burn-up fuels of later campaigns manifest a ratio of less than 1 (as low as 1:10).

In 1970, the first year in which the State of New York determined plutonium ratios in the air of Cattaraugus County, atmospheric plutonium from NFS was clearly detectable beyond the plant confines.[13] The maximum Pu-239 level was 0.00043 pCi/m^3, and the Pu-238 level was 0.00014 pCi/m^3. Assuming an individual breathes $20m^3$ of air per day, and that 10% of the inhaled plutonium is retained in the lungs, the total 50-year dose commitment from these levels of plutonium would be 26.6 and 41.4 mrem, respectively, or a total dose of 68 mrem to the hilar nodes for the 1970 commitment alone. Even higher levels were detected in 1971, but the ratios were such that they could not be clearly attributed to West Valley operations. Plutonium levels in deer lung were first determined in 1971. Maximum deer-lung Pu-238 was 0.72 pCi/kg, and Pu-239, 0.03 pCi/kg.[14] Assuming 2000 g of lung, the 50-year dose commitment to the hilar nodes of these deer would be 190 and 15.6 mrem, respectively. Plutonium concentrations in hilar nodes, rather than whole lung, were not determined.

7.21 Plutonium Contamination of Creeks

Plutonium is also present in Buttermilk and

Cattaraugus creeks, where the maximum concentrations were 0.78 pCi/l of Pu-239 and 1.80 pCi/l of Pu-238 in 1971, the first year in which such measurements were reported.[14] Ingested plutonium concentrates mainly in liver and bone in contrast to inhaled plutonium, which remains in the lung. Had levels been sustained over the course of the year, and assuming that the deer drank 1000 liters of water from Buttermilk Creek at these levels, the total dose commitment to bone from Pu-239 would have been 1 mrem. Unfortunately, deer sampling procedures do not include bone. Neither the Sr-90 nor the plutonium concentration in bone have been determined. Such information would be of considerable value. Plutonium levels in fish and other aquatic biota have also not been determined. It has been determined, however, from studies on salt water species, that marine forms may concentrate plutonium to many times the levels in the surrounding water. Sargassum, for example, has a concentration factor of up to 21,000. Concentration factors of 600-1000 have been observed in algae and up to 2500 in various zooplankton. Marine vertebrates such as tuna, bass, and tautog concentrate plutonium in bone over 500 times.[26]

7.22 Composite Maximum Human Radiation Exposure

Table 7.13 summarizes the maximum dose commitments to a hypothetical individual utilizing maximum observed concentrations in various media during the first five years of NFS operation. (As we have stated already, some but not all of the release rates for radioactivity that lead to these concentrations have since been reduced through the actions of the AEC and the State of New York.) This individual is assumed to have consumed 380 liters of milk, 50 kg of deer meat, 40 kg of fish, and 730 liters of drinking water per year, and to have breathed 20m^3 of air per day, all at the maximum observed concentrations of the various radioisotopes. It is also assumed that this individual spent 3 hours a day fishing at Springville Dam, 250 days a year, during fuel dissolution.

If such an individual is thought to be too hypothetical, it should be pointed out that the maximum isotopic levels in fish were derived from maximum concentrations in fish taken from Cattaraugus Creek and the maximum Kr-85 levels from those observed in Ashford rather than those at the property boundary. Had maximum levels from Buttermilk Creek fish and the property boundary been utilized, the dose commitment would be much higher. For example, the maximum concentration of Sr-90 in the flesh of a sucker taken from Buttermilk Creek in 1970 was 14,456 pCi/kg. Had this fish migrated to Cattaraugus Creek, and been caught by a fisherman at Springville Dam, that individual would have received a 50-year dose commitment to his bone of 55 mrem for each pound of the fish that he consumed. The maximum Sr-90 level in a Buttermilk Creek trout was 5400 pCi/kg, equivalent to a bone dose commitment of 20.6 mrem for each pound consumed. The bone of this same trout contained 320,000 pCi of Sr-90/kg. If a few fish bones were ingested along with the flesh, each gram of the fish bone would deliver a dose commitment to the fisherman of 2.7 mrem (a single teaspoon will hold 9 grams of ground bone).

Table 7.13 Radiation dose to composite "maximum" individual exposed to maximum observed concentration of various waste products released from NFS during the first five years of operation.

Nuclide	Target organ	Medium	Maximum concentration	Dose commitment (mrem)	Total dose commitment (mrem)
Sr-90	bone	milk	28 pCi/l	89	
		deer	543 pCi/kg	227	
		fish	1,679 "	564	
Pu	bone	water	? pCi/l	?	870
Cs-137	whole body	milk	120 "	2.7	
		deer	67,000 pCi/kg	200	
		fish	3,600 "	8.6	
Cs-134	whole body	deer	12,313 "	48	
		fish	1,550 "	5	

Table 7.13 Radiation dose to composite "maximum" individual exposed to maximum observed concentration of various waste products released from NFS during the first five years of operation (continued).

Nuclide	Target organ	Medium	Maximum concentration	Dose commitment (mrem)	Total dose commitment (mrem)
Tritium	whole body	milk	2,300 pCi/l	0.3	
		deer	683,550 pCi/kg	4.4	
		fish	92,255 "	1.0	
		air	0.12 pCi/l	0.13	
Sediments	whole body	bank of Cattaraugus Creek	0.020 mR/hr	15	
Kr-85	whole body	air	7.8 pCi/m^3	1.4	286.5
Tritium	whole body	drinking water	Cattaraugus Creek 184,510 pCi/l	17.5	
I-129	thyroid	milk	1.2 "	2.7	

The maximum Sr-90 content of the bone of a Buttermilk Creek sucker in 1970 was 620,000 pCi/kg, equivalent to 5.2 mrem per gram. It would have behooved the fisherman in the vicinity of NFS to have filleted his fish carefully.

The total bone dose commitment to the maximum individual, including Sr-90 from milk, deer and fish, would have been 1156.5 mrem (see Table 7.13 - 870 millirems to bone and 286.5 whole body). The dose to bone from plutonium is unknown. Seven percent, or 61 mrem, of the Sr-90 dose would have been delivered in the first year. In addition to an unknown dose from the plutonium isotopes, the commitment from the previous year's Sr-90 ingestion, other bone-seeking nuclides, Sr-90 in fish bones, and the bone dose from whole body irradiation would easily have exceeded 350 millirems, could this dose be determined accurately. If, as suggested in the National Academy of Science-National Research Council report, the emissions from nuclear facilities were limited to a fraction of the total population radiation exposure - say 10% - the predicted dose to the hypothetical maximum individual would have exceeded the limit.

The postulated maximum individual would have accrued a whole body dose of 286.5 mrem, or greater than 50% of the current limit, from consumption of milk, deer, fish containing observed maximum levels of Cs-134 and Cs-137 and tritium, as well as from drinking tritiated water, and breathing air containing tritium and Kr-85. Had the maximum fish levels from Buttermilk Creek been utilized, these doses would have been considerably higher. Although tritium levels in the public water supplies are much lower than in the creeks, and produce an annual dose of less than 1 mrem, should the hypothetical maximum individual have chosen to derive his entire drinking water supply from Cattaraugus Creek at its maximum tritium level, he would have accrued an additional whole body dose of 17.5 mrem per year. It has also been assumed that this individual spent 750 hours along the banks of Cattaraugus Creek (with an average creek background of 0.020 mrems per hour rather than the maximum observed level of 0.132 mrem per hour) and with "maximum" atmospheric tritium and Kr-85 levels.

7.23 Composite Maximum Deer Radiation Exposure

The maximum hypothetical deer would receive almost 300 mrems whole body per year (Table 7.14), assuming he drank 1000 liters of Buttermilk Creek at maximum tritium levels and inhaled 20 m^3 of air containing Kr-85 at its "maximum" observed level in Ashford. The tritium contribution would be 8.9 mrem per year; Cs-137, 120.5; Cs-134, 29.6; Sr-90, 136.9; and Kr-85, 1.4. At the maximum observed thyroid I-129 concentration, the thyroid dose would be 5.6 rem per year, which exceeds the maximum human limit of 1.5 rem by a factor of 3.7. The bone dose, from plutonium at maximum observed levels in Buttermilk Creek, would be approximately 1 mrem from Pu-239. The bone dose from Sr-90 ingestion at maximum Buttermilk Creek levels could be up to 16.7 millirems, for a total bone dose commitment of 16.8 mrems per year. Assuming 7% of the Sr-90 and 2% of the plutonium dose commitment is delivered in the first year, the annual bone dose would be approximately 1176 mrem, or 2.8 times the maximum bone dose permitted humans. Un-

Table 7.14 Radiation dose to composite "maximum" deer exposed to and containing maximum observed concentrations of various waste products released from NFS during the first five years of operation.

Nuclide	Target organ	Medium	Maximum concentration	Dose commitment (mrem)	Total dose commitment (mrem)
Sr-90	bone	bone	unknown pCi/kg	?	
Pu-238	"	"	"	?	
Pu-239	"	"	"	?	
Sr-90	whole body	muscle	543 pCi/kg	136.8	
Cs-137	"	"	66,956	120	
Cs-134	"	"	12,313	29.5	
Tritium	"	"	683,550	8.8	
Tritium	"	air	0.12 pCi/m^3	0.13	

Table 7.14 Radiation dose to composite "maximum" deer exposed to and containing maximum observed concentrations of various waste products released from NFS during the first five years of operation (continued).

Nuclide	Target organ	Medium	Maximum concentration		Dose commitment (mrem)	Total dose commitment (mrem)
Kr-85	whole body	air	7.8	pCi/m^3	1.4	
Sediments	"	creek banks	?		15	311.6
I-129	thyroid	thyroid	3700	pCi/g	5,600	
Pu-238	hilar nodes	lung	0.72	pCi/kg	190	
Pu-239	"	"	0.03	"	15.6	

fortunately, deer taken from the NFS site have not been analyzed for bone Sr-90 and plutonium content.

It is apparent that the failure to include the wildlife in the radiation exposure considerations has resulted in substantial irradiation of deer and probably other denizens of the woods around NFS. Without compelling reasons to the contrary, it is illogical not to apply the same standards to wildlife as to humans. The fallacy of such an oversight should be self-explanatory and is strongly reinforced by the excessive thyroid and bone doses calculated for local deer. Although cesium and strontium levels declined during 1970 and 1971, they can be expected to increase in the future. The estimated maximum exposures may well be exceeded in the future if the standards are not applied to wildlife. I-129 levels will certainly increase unless stack discharges are controlled. Tritium levels, and resulting whole body exposure, will increase inexorably.

References

1 Hearings before the Joint Committee on Atomic Energy, Congress of the United States, Part II, Vol. 1, 1970.

2 "An Investigation of Airborne Radioactive Effluent from an Operating Nuclear Fuel Reprocessing Plant," BHR/NEHRL 70-3.

3 "An Estimate of Radiation Doses Received by Individuals Living in the Vicinity of a Nuclear Fuel Reprocessing Plant in 1968," BRN/NERHL 70-1.

4 State of New York, Department of Environmental Conservation, "1969 Annual Report of Environmental Radiation in New York State," Dec. 1, 1970.

5 "A Review of the Radiological and Environmental Aspects of Krypton-85," NF-69-16.

6 J.L. Russell and P.B. Hahn. "Public Health Aspects of I-129 from the Nuclear Power Industry," Rad. Data and Health Reports April 1971, pp. 189-194.

7 P.M. Bryant. "Derivation of Working Limits for Continuous Release Rates of I-129 to the Atmosphere," Health Phys. 19 611-619 (1970).

8 J.C. Daly, A.V. Manchester, J.J. Gabay, and N.I. Sax. "Tritiated Moisture in the Atmosphere Surrounding a Nuclear Fuel Reprocessing Plant," Rad. Data and Health Reports 1968, pp. 341-346.

9 Hearings before the Joint Committee on Atomic Energy, Congress of the United States, Part III, 1971.

10 "Liquid Waste Effluents from a Nuclear Fuel Reprocessing Plant," BRH/NERHL 70-2, 1970.

11 Hearings before the Joint Committee on Atomic Energy, Congress of the United States, Part I, 1969.

12 State of New York, Department of Environmental Conservation. "1968 Annual Report of Environmental Radiation in New York State."

13 State of New York, Department of Environmental Conservation. "1970 Annual Report of Environmental Radiation in New York State."

14 State of New York, Department of Environmental Conservation. "1971 Annual Report of Environmental Radiation in New York State."

15 State of New York, Department of Environmental Conservation. "Radioactivity in Air, Milk, and Water for January-March 1972."

16 State of New York, Department of Environmental Conservation. "Radioactivity in Air, Milk, and Water, April-June 1972."

17 NCRP Report #39, "Basic Radiation Protection Criteria," 1971.

18 B. Schlien. "An Evaluation of Internal Radiation Exposure Based on Dose Commitments from Radionuclides in Milk, Food, and Air,"

Health Phys. 18 267-276 (1970).

19 C.L. Weaver, E.D. Harward, and H.T. Peterson. "Tritium in the Environment from Nuclear Power Plants," Rad. Data and Health Reports 84 363-371 (1969).

20 S.M. Jinks and M. Eisenbud. "Concentration Factors in the Aquatic Environment," Rad. Data and Health Reports May 1972, pp. 243-247.

21 "The Effects on Populations of Exposure to Low Levels of Ionizing Radiation," Report of the Advisory Committee on the Biological Effects of Ionizing Radiations (BEIR Report), Division of Medical Sciences, National Academy of Sciences, National Research Council, November 1972.

22 N.L. Sax, P.C. Lemon, and J.J. Gabay. "Radioecological Surveillance of the Waterways Around a Nuclear Fuels Reprocessing Plant," Rad. Data and Health Reports 1969, pp. 289-296.

23 F.W. Whicker. "Radioecology of Some Natural Organisms and Systems in Colorado," Colorado State University, April 1, 1972, COO-1156-54.

24 D.G. Jacobs. "Sources of Tritium and Its Behavior Upon Release to the Environment," Appendix 5 in Hearings before the Joint Committee on Atomic Energy, Congress of the United States, Part I, 1969.

25 J.A. Norcross and C.E. Newton, Jr. "U.S.

Transuranium Registry, A Progress Report." Health Phys. 22 887-890 (1972).

26 V.E. Noshkin. "Ecological Aspects of Plutonium Dissemination in Aquatic Environments." Health Phys. 22 537-549 (1972).

27 J.A. Martin. "Calculations of Environmental Radiation Exposures and Population Doses Due to Effluents from a Nuclear Fuel Reprocessing Plant." Rad. Data and Health Reports 14, No. 2 (February 1973).

Chapter 8

STORAGE AND DISPOSAL OF HIGH-LEVEL RADIOACTIVE WASTES

Thomas C. Hollocher

8.1 Summary and Recommendations

Except for the storage of liquid wastes in tanks, for which experience from weapons production applies, all proposals for long term storage or disposal of high-level waste from the nuclear power industry lie at the research and development stage. The proposals so far considered seriously by the AEC for the disposal of wastes are dubious in concept (caverns excavated by nuclear weapons, depositing wastes in liquid form in rock caverns), not technically feasible (disposal in solar orbit), or they are so dependent upon site-specific geological characteristics that suitability cannot be determined a priori without extensive on-site investigation (disposal in bedded salt or under Antarctic ice). One such site, the salt beds at Lyons, Kansas, was studied carefully for several years in the 1960s and was finally adjudged unsuitable. The matter of man-made vaults at or near the surface for long-term storage (500 years) is equally uncertain. Their maintenance depends upon the existence of social and political institutions, the permanence of which cannot be guaranteed over time.

The impression is inescapable, in view of the present imprecise state of affairs, that no convincing statements exist regarding the long-term environmental impact attending the storage and/or disposal of wastes from fuel reprocessing. It is disturbing in this regard that the pace of development of a suitable waste disposal technology does not seem commensurate with the rate of expansion of the nuclear power industry. This situation is reminiscent of developments attending the production of plutonium for the weapons program: hasty and, in part, incorrect decisions were made regarding the storage of wastes produced by a technology alleged to be urgently necessary.

The times during which radioactive wastes must remain secure from the biosphere have no parallel in human affairs. Eight hundred years are required for fission products alone and millions of years if the fission products continue to be contaminated with transuranic elements at present levels. Fission technology requires that man issue guarantees on events far into the future, and it is not clear in most cases how this can be done. Institutional arrangements do not exist and never have existed to guarantee the monitoring of or attendance upon storage facilities over a millennium. In the range of a million years, serious geological uncertainties arise and even the survival of man may be doubtful. "In perpetuity" has little real meaning in human affairs.

The uncertain state of affairs in high level waste disposal was well summarized in 1972 by F.K. Pittman, director, Division of Management and Transportation, USAEC.[29]:

"While none of the suggested long-term solutions to the problem of permanent disposal of high-level radioactive waste is technically or economically feasible today, the AEC recognizes that one or more may well offer attractive options in the future."

It is interesting to note that AEC regulations now stipulate that high-level wastes are to be solidified after 5 years' storage and shipped to a federal repository within 10 years. The repository does not now exist and its final form is uncertain. In the meanwhile, billions of curies from commercial power production will join the accumulated several billion curies of weapons waste - all of it at the earth's surface and vulnerable to accident, error, or sabotage.

It is probably advisable, given the scale of anticipated nuclear operations, to increase the budget for research and development in waste management. Through the 1960s to 1971, the AEC budget for this category was about $5 million/yr. Of that, about $1/2 million/yr went to salt disposal, specifically the Lyons, Kansas project, about $2 million/yr was committed to waste solidification technology, and about $2.5 million/yr has been used to develop a rock cavern disposal scheme for the Savannah River plant. This $5 million can be compared with a total yearly waste management budget in the AEC of about $35 million, most of which goes to management of weapons wastes. For orientation, the entire AEC budget is of the order $2.5-3 billion/yr.

8.2 Specific Recommendations

1. The chemistry of fuel reprocessing should be improved to separate almost totally the long-lived transuranium elements and other actinides from fission products. Levels of actinides should be reduced by 10-fold for some, and 100-fold for others relative to

present levels in the reprocessing waste stream. The extraction of uranium, neptunium, and plutonium should be 99.95% complete and that of americium and higher actinides should be 99% complete. This will result in increases in the cost of fuel reprocessing and, together with Recommendation 2 below, lead to an increase of about 2-3% in the cost of delivered nuclear power. At present, for comparison, the only actinides recovered are uranium and plutonium, both at the 99.5% level.

Similar improvements should be made also in cleaning up fuel cladding hulls and alpha (α) wastes from fuel fabrication.

2. Recovered transuranic elements should be disposed of through their transmutation and fission, preferably in fast neutron reactors. This would diminish the time of hazard for the remaining fission products to something of the order of 800 years, as opposed to millions of years otherwise.

3. The remaining fission products should be solidified in a time as short as is consistent with radiogenic heat production, and certainly no greater than 5 years, and be placed, as soon as is practicable and safe, in carefully selected geological formations. The possibility that solidification can immediately follow reprocessing of fuel should be vigorously pursued, for this would eliminate interim tank storage of liquid wastes.

4. Long-term (> 50 year) storage of fission products in surface or near surface facilities should not be undertaken.

5. Disposal plans which are attractive in principle but at present unfeasible (disposal in solar orbit) or in need of further study (disposal under the Antarctic ice cap) should

be reviewed periodically in light of relevant technological developments or new data.

6. In light of the impending need for a federal waste repository, a greatly increased research and development effort toward that end should be undertaken by the AEC and the relevant industrial authorities. A detailed plan for the disposal of high-level wastes through 2000-2010 A.D. ought to appear within the next few years.

8.3 High-Level Radioactive Wastes and Their Decay Characteristics

During the operations of a reactor, two main types of radioactive products accumulate in the fuel: fission products and actinides, the most abundant of the latter being the transuranium elements. Actinides are the series of elements actinium (Ac), thorium (Th), protactinium (Pa), uranium (U), and the transuranium elements starting with neptunium (Np), plutonium (Pu), americium (Am), and curium (Cm). Fission products, as the name implies, are the products of the fission of U-234 and, to some extent in light water reactors, of Pu-239 and Pu-241, and have mass numbers in the range 80-155. The transuranic elements are built up by neutron capture, starting with U-238, and include fissionable and fertile isotopes of plutonium which are of value as nuclear fuel. The transuranic elements of major importance here have mass numbers from 237 to 244.

In the irradiation of enriched uranium fuel in a light water reactor to the level of 33 x 10^9 watt-days (thermal) per metric ton of fuel, about 33 kg of uranium are consumed (converted to fission products) per metric ton and there is present at the end 8-9 kg of plutonium and about 1 kg of other transuranic actinides, most of which is neptunium.[1]

During fuel reprocessing the fission products are separated, and uranium and plutonium are recovered. The separations are not quantitative (i.e., total) under present chemistry (Purex and related processes) and about 0.5% of the plutonium and uranium and most of the other actinides appear with the fission products in the final aqueous waste stream. Typical values for the yields of fission products and transuranic elements in the reprocessing waste stream are given in Tables 8.1 and 8.2, respectively. In addition, about 0.1% of the plutonium remains with the fuel cladding hulls. Some also appear as alpha wastes in fuel fabrication.

It is clear that both the short-lived products with half lives less than one year and the very-long-lived isotopes with half lives greater than 10 million years are of little concern in the context of long-term storage and disposal. The former decay largely within the period of thermal cooling of 3 to 5 years which can be considered temporary storage; the latter decay so slowly that they exhibit low specific activities and rather low radiological hazards. Iodine-129 with a half-life of 17 million years, for example, has a specific activity of only 0.16 mCi/g and would not represent a major biological hazard[2] even to thyroid tissue if it were diluted with stable iodine to the level of 1 part in 10^5-10^6. In Tables 8.1 and 8.2, therefore, we have considered only products with half lives greater than one year and (except for I-129) less than 10 million years.

The high initial radioactivity of fission products is occasioned by short-lived isotopes. If total decay energy of fission products at age 1 day were taken as 100, then the value would drop to 10 at 100 days, to 1 at about 900 days, to 0.3 at 10 years, and to 0.1 at about 41 years.[3] At 5 years, the total decay energy is only about twice that for Sr-90 and Cs-137 combined, and, as seen in Table 8.1, these two isotopes contribute the bulk of the decay energy at ten years. By 500 years, the decay energy has dropped another 4 orders of magnitude relative to that at 10 years and is now dominated by SM-151. A further factor of decrease of about 2 is realized by 1000 years at which time Zr-93, Tc-99, and Cs-135 dominate. Thereafter decay is very slow and is governed by isotopes with half lives greater than 2×10^5 years. In summary, fission product decay energy drops by a factor of several million after 800 years relative to the value at 1 day.

While fission products are not without hazard after 800 years, the relative toxicity is only about 50 times that of a similar volume of natural uranium ore containing 1.4% U_3O_8, by weight.[4]

We define relative toxicity at different times according to the approach of Kubo and Rose.[4] For each radioactive species, a water dilution factor is computed so that the species would be present at no more than the maximum permissible concentration set by national standards. The dilution factors, summed over all the species, are then compared with the dilution factor for uranium ore of the above type, which serves as a nonzero but commonly encountered source of radioactive

Table 8.1 Typical yields of fission products with intermediate and long half lives from processing one metric ton of light water reactor fuel irradiated to 33 x 10^9 watt-days (thermal) per metric ton of fuel.[a]

Element	Half-life	Curies remaining after decay of			
		10 yr	100 yr	500 yr	1000 yr
Ce-144/Pr-144	285 day	300	--	--	--
Ru-106/Rh-106	367 day	1100	--	--	--
Eu-155	1.8 yr	160	--	--	--
Cs-134	2.1 yr	8300	--	--	--
Sb-125/Te-125m	2.7 yr	980	--	--	--
Sr-90/Y-90	28.1 yr	1.2 x 10^5	1.32 x 10^4	0.6	--
Cs-137/Ba-137a	30 yr	1.6 x 10^5	2.1 x 10^4	2	--
Sm-151	90 yr	1100	520	30	0.4

Table 8.1 Typical yields of fission products with intermediate and long half lives from processing one metric ton of light water reactor fuel irradiated to 33 x 10^9 watt-days (thermal) per metric ton of fuel (continued).

Element	Half-life	Curies remaining after decay of			
		10 yr	100 yr	500 yr	1000 yr
Tc-99	2.1 x 10^5 yr	15	15	15	15
Zr-93	9 x 10^5 yr	3.7	3.7	3.7	3.7
Cs-135	2 x 10^6 yr	1.7	1.7	1.7	1.7
Pd-107	7 x 10^6 yr	0.013	0.013	0.013	0.013
I-129	17 x 10^6 yr	0.025	0.025	0.025	0.025
Total Curies (approx.)		300,000	35,000	53	22

a. Calculated from data in refs. cited in notes 1, 7, and 28. A 1000 Gw(e) plant will yield about 30 tons of fuel for reprocessing per year.

Table 8.2 Typical yields of actinides with stream from processing one metric ton of light days (thermal) per metric ton of fuel.[a]

Amount Present After Decay of:					
Element	Half life	Reprocessing 0.3 yr		10 yr	
		g	Ci	g	Ci
Np-237	2.1×10^6 yr	760	0.59	760	0.59
Pu-238	86 yr	5.8	105	5.5	100
Pu-239	24,400 yr	27.5	1.7	27.5	1.7
Pu-240	6,580 yr	8.5	2.0	19.2	4.5
Pu-241	13.2 yr	4	464	2.4	273
Pu-242	379,000 yr	2	0.009	2	0.009
Am-241	462 yr	54	189	55.6	198
Am-243	7,370 yr	82	17.0	82	17.0
Cm-244	17.6 yr	30	2570	19.7	1700
Total g/metric ton fuel		974		974	
Activities, including daughter products (approx.)			200,000		10,000

a. Calculated from data in refs. cited in yield about 30 tons of fuel for reprocessing intermediate and long half lives in the waste water reactor fuel irradiated to $33x10^9$ watt-

500 yr		10,000 yr		100,000 yr	
g	Ci	g	Ci	g	Ci
786	0.61	810	0.63	790	0.61
0.1	1.8	--	--	--	--
32	2.0	59	3.6	4.5	0.3
38.4	8.8	13.9	3.2	--	--
--	--	--	--	--	--
2	0.009	2	0.009	1.7	0.007
29.5	103	--	--	--	--
77	16.0	31	6.5	--	~ 0.001
--	--	--	--	--	--
965		916		796	
	900		100		50

notes 1, 7, and 28. A 100 Gw(e) plant will per year.

isotopes in the lithosphere. The reprocessing wastes are assumed to be solidified at 1 ft^3 per 10 x 10^9 watt-day (thermal operation of the reactor. This approach takes into account both the direct radiological hazards and the biological factors, such as body turnover and organ specific sequestration, that are believed to apply in men. It does not consider more complicated factors, such as concentration mechanisms which often apply to isotopes as they pass through a series of ecological food chains.

The problem with the transuranic elements is more difficult. Elements of the group decay by alpha (α)- and/or beta (β)-emission through numerous steps into other radioactive daughter elements (e.g., Cm-244 into Pu-240); and they have generally very high biological toxicity by virtue of α-emission and biological sequestration. Three disintegration processes are of importance in Table 8.2:

$$\text{Cm-244} \xrightarrow[\alpha]{17.6\text{ yr}} \text{Pu-240} \xrightarrow[\alpha]{6580\text{ yr}} \text{U-236}.$$

$$\text{Am-243} \xrightarrow[\alpha]{7370\text{ yr}} \text{Np-239} \xrightarrow[\beta^-]{2.35\text{ day}}$$

$$\text{Pu-239} \xrightarrow[\alpha]{24{,}400\text{ yr}} \text{U-235}.$$

$$\text{Pu-241} \xrightarrow[\beta^-]{13.2\text{ yr}} \text{Am-241} \xrightarrow[\alpha]{462\text{ yr}}$$

$$\text{Np-237} \xrightarrow[\alpha]{2.2\text{x}10^6} \text{Pa-233}.$$

The second series of decays generates Pu-239 and is particularly relevant to very-long-term storage. One decay path of long-term interest not shown in Table 8.2 is the decay of Pu-238 into U-234 which itself has a half life of 1/4 million years.

Plutonium is perhaps the most toxic element known, and represents a chronic biological hazard on a weight basis comparable to the acute hazard from botulinus toxin, i.e., it is hazardous to 10^{-6} g, if appropriately deposited in lungs or bone.[5] Plutonium-239 in equilibrium with its daughters has a direct radiological hazard about 10,000 times that of natural uranium in equilibrium with its daughters.[6] Plutonium exhibits, in addition, a very large biological concentration factor with little turnover in bony tissues. The yields of Pu-239 and Pu-240 from one metric ton of fuel in Table 8.2 represent together about 70 g of plutonium after 500 years' storage. It is clear that the presence of current levels of transuranic elements alone in the waste stream will require that the wastes be secured from the biosphere for very long periods.

The period of hazard for the waste actinides is determined in the long term by the properties of the radioactive daughters. The decay energies of the actinides and their daughters decrease with time and become equal to those of fission products after a few hundred years, as shown in Figure 8.1. Thereafter the actinides dominate. The relative toxicities of the decay daughters tend, however, to be somewhat greater than the parent actinides.[4] As seen in Figure 8.2, the relative toxicity goes through a minimum at about 2×10^4 years and

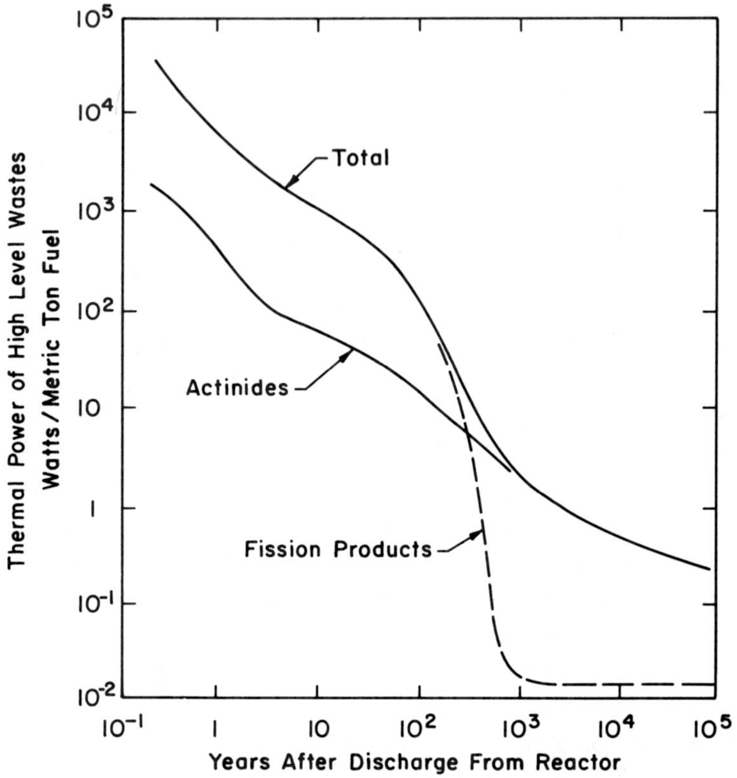

Figure 8.1 Thermal power from light water reactor wastes. Fuel utilization is taken to be 33 x 10^9 watt-days (thermal) per metric ton with recovery of uranium and plutonium only with 99.5% efficiency. Adapted from ref. cited in note 4.

then increases several-fold and reaches a maximum at about 7 x 10^5 years. Relative toxicity remains more than 1000 times that of uranium ore over most of 4 x 10^6 years.

8.4 Growth of Nuclear Power

An example of a projection on the growth of the nuclear power industry in the United States is given in Table 8.3. Various pro-

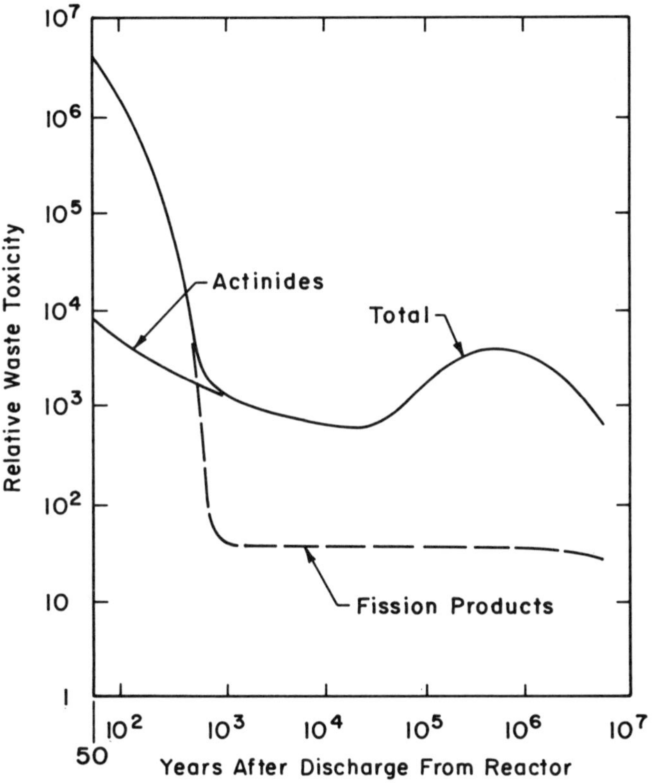

Figure 8.2 Relative toxicity of wastes from light water reactors. Burn-up and recoveries are given in Figure 8.1. The relative toxicity is defined in the text and is relative to uranium ore. Adapted from ref. cited in note 4.

jections have been made and, while some disagreement is noted, the projections generally agree within a factor of about 2.[1,6,7,8]

It seems likely that 700-1000 reactors will be in operation by the year 2000, each with an output of about 1 Gw(e) (1000 megawatts) or 8000 Gw-hrs(e)/hr. This is a level of about 35 times the output of the present nuclear power industry. Each reactor will

Table 8.3 Projected waste inventories at the permanent repository from nuclear power generation.[a]

	By the Calendar Year		
	1980	1990	2000
Installed nuclear capacity (Gw(e))	100	420	734
Accumulated high-level waste			
Vol. of waste (10^3 ft^3)	3.2	74.2	319
Thermal output (10^6 w)	1.2	24.4	94.9
Total radioactivity (x 10^6 Ci)	329	7020	27700
Sr-90 (x 10^6 Ci)	59.0	1310	5290
Cs-137 "	83.1	1850	7500
Pu-238 "	0.102	2.34	9.88
Pu-239 "	0.002	0.04	0.16
Pu-240 "	0.004	0.10	0.47

Table 8.3 Projected waste inventories at the permanent repository from nuclear power generation (continued).

	By the Calendar Year 1980	1990	2000
Am-243 (x 10^6 Ci)	0.15	3.54	15.3
Cm-244 "	1.58	34.1	133.3
Accumulated alpha waste			
Thermal output (10^6 w)	0.014	0.17	0.48
Total radioactivity (10^6 Ci)	14.2	151	300
Total mass actinides (10^6 g)	1.4	15.8	38.3
Pu-238 (10^6 Ci)	0.23	2.57	6.02
Pu-239 "	0.05	0.58	1.41
Pu-240 "	0.07	0.83	2.02
Pu-241 "	13.8	146	286

Table 8.3 Projected waste inventories at the permanent repository from nuclear power generation (continued).

	By the Calendar Year 1980	1990	2000
Am-243 (10^6 Ci)	0.06	1.03	4.74

a. From refs. cited in notes 6, 7, and 8. Half the waste is assumed to have aged 5 years and half 10 years.

generate 5-8 metric tons of solidified high-level waste per year at steady state, to yield a total of about 5000 metric tons of waste per year. Accumulated wastes will be about 25,000 metric tons by the year 2000, or, if left as aqueous liquid waste, 40-60 million gallons. Accumulated radioactivity has been estimated to lie in the range of 200 billion Ci, with about 30 billion Ci residing in high-level waste which has aged for 5-10 years. The accumulated Pu-239 alone would amount to about 24 metric tons total in the waste. The capital investment for the plants alone is estimated to be about $300 billion. Sales of nuclear-generated electricity at 20-25 mills per kw-hr will result in a $160-200 billion/yr industry.

8.5 Intermediate and Low-Level Wastes

A second class of nuclear waste exists which contains radioactivity at much lower levels - the so-called low- and intermediate-level wastes.[1,7,9] These are comprised largely of solutions and materials which have become contaminated with various but generally small amounts of fission products and/or transuranic elements as the result of processes or accidents within the nuclear industry. An important source of these wastes in the power industry will be the fuel cladding hulls which contain activation elements and some residual plutonium. The problem of disposal of low- and intermediate-level wastes lies not so much in the amounts of radioactivity, which are small compared to those arising directly from fuel reprocessing, but rather in their large volume and varied form.

Low- and intermediate-level wastes have

been disposed of variously since the onset of nuclear weapons production: direct release to surface waters or discharge into the ground by seepage in the case of aqueous solutions, and shallow burial of solids in both recoverable and nonrecoverable forms. The chief problems concern the reduction of bulk, which, for some material, can be accomplished by combustion and fluxing of the solid residue at high temperatures.

While not perhaps a major problem, it is clear that a large nuclear power industry could deposit potentially large and serious amounts of radioactive materials in the form of low- and intermediate-level wastes. It is important that the disposal of these wastes be very carefully considered. Some lower-level wastes can be reduced in volume and ultimately added to high-level wastes for long term storage or disposal. For other materials, separate disposal facilities may be in order. Fuel cladding hulls may represent a special problem in the long run. About 1/6 of the unrecovered plutonium (8-9 g plutonium per metric ton of fuel reprocessed) remains with the hulls.[1,7] It may prove advisable or necessary to make the recovery of transuranic elements from the hulls nearly complete.

8.6 Experience With Wastes From Nuclear Weapons Production

Since the inception of the nuclear weapons program during World War II, men have had to deal with the problem of handling and storing high-level nuclear wastes. There is now in storage from the weapons program about 90 million gallons of liquid waste and an accumulated total of several billion Ci of fission products.[9] The material must be considered to be under temporary storage. In retrospect, it is now clear that mistakes were made early in the technology of storage of high-level wastes arising from the weapons program. It is interesting that the mistakes were occasioned, in part at least, by the fact that hasty decisions and shortcuts were forced by what seemed to be the overriding national need for bomb-quality plutonium. It is not necessarily true that all facets of a new technology can be forced into efficient operation at the same rate. The lessons learned as the result of weapons technology should be carefully applied in the nuclear power industry.

High-level fission products from fuel reprocessing are thermally hot initially and are stored temporarily in water which is cooled either by refluxing or by refrigeration. The products partly dissolve and partly form slurries depending on their solution chemistries and pH. After 3 to 5 years, the rate of heat production has dropped to the point where the water can be removed and other long-term storage or disposal strategies can be considered.

The bulk of wastes from nuclear weapons production are stored as neutral or alkaline solutions/slurries in underground carbon steel tanks, with lesser amounts stored as acid solutions in stainless steel tanks, especially at the Idaho Chemical Processing Plant. The Hanford, Washington, and the Savannah River, South Carolina, facilities house, between them, most of the high-level liquid wastes.[1,9]

The approximately 90 million gallons of waste are stored in some 200 tanks with capacities between 0.3 and 1.3 million gallons, encased in concrete and surrounded by earth under shallow burial. The capital costs for the tanks range between $0.50/gallon for 1.3 million gallon carbon steel tanks without cooling facilities, to $5.40/gallon for 0.3 million gallon stainless steel tanks with cooling coils and agitators.[1]

While the tanks had life expectancies in excess of 50 years, the combination of heat, radiation, high salt concentrations, and mechanical stresses (especially at welds) has resulted in tank failures starting in 1958 and unexpectedly short operational lives. As of 1970, 15 tank failures had been reported, all in carbon steel tanks at Hanford and Savannah River.[1] By 1973, the 16 failures reported at Hanford alone have released an estimated 230,000 Ci of Cs-137 to the ground beneath the tanks. (See Appendix 8.1 for a description of a recent large leak at Hanford.) The total released, including Sr-90, is about 350,000 Ci. Most of the radioactivity from the smaller spills is believed to be retained by the soil 3-5 m below the tanks, and within 30 m in the case of larger leaks. A total of 4 lesser failures have occurred at Savannah River. In addition, one release was reported at Savannah River involving about 2000 Ci of Cs-137 during an accident in a transfer operation. The waste was released into a sewer and thence partly into a connecting stream. No leaks have been reported in the stainless steel tanks at the Idaho plant over 16 years, but severe pitting and weld corrosion has been noted. A number of underground leaks in transfer pipelines have occurred at Hanford.

In the case of tank damage, the remaining contents are pumped to another tank and the damaged tank is withdrawn from operation or sealed. Stainless steel tanks can be stripped of residual solids and scale by acid washes, and therefore can be largely decontaminated. Carbon steel cannot easily be cleaned by acid, with which it reacts. Thus, satisfactory methods may not exist for decontaminating carbon steel tanks. A combination of materials scarcity and cost resulted in the construction of most tanks from carbon steel rather than from stainless steel. This is in retrospect a serious mistake involving considerable hazard and additional major costs which have yet to be evaluated.

The amounts of radioactivity which have already or may eventually reach the biosphere, via leaching and via uptake by plant roots, earthworms, and burrowing insect larvae, from these spills are not well known. It is known, however, that the Hanford facility is a major cause of the radioactive contamination of the Columbia River and that wildlife at the site, including ducks, have exhibited significant contamination in recent years. Most probably this is the result more of the routine practice of discharging low- and intermediate-level wastes into the ground than of the spills per se. Hanford is located in a rather arid region which exhibits a deep "dry soil" zone between the surface and the water table. It was long assumed, therefore, that radioactivity injected into surface soil layers would not reach the water table and undergo rapid migration. The belief has been questioned, however, that the "dry soil" zone is penetrated neither by downward percolation of rain water or snow melt, nor by upward migration from the water table into the root zone.[10] In fact, the latter situation may operate to return radioactive elements to the surface.

Actual evapotranspiration often appears to exceed rainfall at Hanford. These are only some of the major objections to the emplacement of wastes into the ground at Hanford.[10]

In this regard, a situation recently has come to light[11] which emphasizes my belief that the previous handling of radioactive wastes has involved careless and potentially dangerous procedures. Perhaps 300 kg of plutonium from Hanford's Plutonium Finishing Plant have been allowed to leach into the ground under concrete-covered trenches. The description of the situation at trench Z-9 given in Appendix 8.2 is taken directly from the AEC report, WASH-1520.[11] Note that the fourth paragraph of Appendix 8.2 suggests that, under some circumstances, a nuclear chain reaction could occur in this trench. Trench Z-9 contains coil contaminated with an estimated 100 kg of plutonium, which are thought to be 75% recoverable and valued at $3 million. The Natural Resources Defense Council, Inc. has information[12] indicating that 13 other seepage cribs or trenches at Hanford may contain an additional 200 kg of plutonium. The element is alleged to have remained absorbed in the upper levels of soil, but the available data on this point leave much to be desired. As suggested above, rather direct routes of transfer of radioactive elements from ground to the biosphere are known.

In the event of a plutonium chain reaction, trench Z-9 might suffer a small nuclear excursion of sufficient force to throw tons of radioactive particles into the air. Contamination far beyond the Hanford site would be likely.

Substantial amounts of wastes have also

been disposed of intentionally by the AEC in drainage pits and trenches in areas of heavy rainfall. At Oak Ridge National Laboratory, Tennessee, for example, a total of more than 1 million Ci of locally generated liquid intermediate wastes were poured into the ground between 1952 and 1966, including about 280,000 Ci of Sr-90. After termination of use, the pits and trenches were asphalt-covered to minimize infiltration of rainwater, but this remedy offers only short-term and partial protection at best.

The total AEC waste management budget in the 1970s, exclusive of research and development programs, has been running about $35 million per year. The management of nuclear weapons wastes now stored in tanks may in the long run cost an additional $1 billion.

Programs are now underway, but have not proceeded very far, to convert weapons waste from liquid form to other forms and thus to phase out tank storage. At Hanford, the main heat producing isotopes, Sr-90 and Cs-137, are being stripped out, encapsulated, and stored under water. The remaining wastes are then concentrated or solidified to salt cake. This program is connected to another proposal to deposit the salt cake or dense slurry into caverns in the basalt underlying the site. Environmental objections to this plan have been major and its future is very much in doubt.[13] A similar plan to deposit dense slurries in the gneiss and sedimentary formations under the Savannah River plant has been under study since 1960. At the Idaho plant, wastes are being calcined to granular solids and stored in stainless steel bins in underground concrete vaults to await final disposal. Of about 4 million gallons of liquid waste generated at the Idaho plant, about 2 million gallons have been converted to solid form over the past ten years.

It is astonishing, therefore, in view of the deficiencies in tank storage and in view of an avowed intent to convert liquid wastes to solid form, that the AEC plans now to build 4 additional carbon steel tanks, each of 1.3 million gal. capacity, for the future high-level waste facility at Savannah River.[14]

8.7 Storage and Disposal Proposals in the Nuclear Power Industry

Within a decade, accumulated wastes from nuclear power generation will be in the billion-Ci range and reach tens of billions of Ci by the turn of the century. It is clear that indefinite tank storage is unsatisfactory and expensive; on the other hand, proposals for alternate long-term storage programs and for permanent disposal methods are still in the research and development state. Waste storage and disposal represent certainly two of the most crucial problems facing the nuclear power industry. Incorrect decisions in this decade could lead to enormous expenses in the next two decades and possibly even to a crippling of the nuclear power industry.

The billions of Ci of high level wastes expected over the next 30 years must not under any circumstances be released into the biosphere. At present only three disposal options seem open: (a) the wastes may be removed from the earth, (b) they may be contained on or in the earth for sufficient periods - 800 years for fission products and millions of years for transuranic actinides - and, (c) the transuranic actinides may be transmuted to fissionable isotopes and destroyed by fission. In the following we shall evaluate these various proposals for storage and/or disposal. For the purposes of definition, "storage" implies retrievability, i.e., mistakes can be corrected in principle; "disposal" implies irretrievability. In passing I should mention that transmutation of fission products, as opposed to transuranic elements, to stable isotopes is entirely impractical.

Storage and disposal (in the earth) are two very different stratagems. Storage retains future options and provides safety against error, but requires full societal responsibility. Disposal guards against an irresponsible society but offers the possibility of environmental hazard at some future time. Decisions regarding storage and disposal are by no means entirely technological. They are to a large extent societal in nature and relate to very uncertain matters: the stability of societies over time; the extent of responsibility we have to future generations of man and even to the biosphere generally. In addition, there is a desirable economic goal that waste management in the power industry should cost only a small fraction of the cost of delivered electricity.

8.7.1 Short-term storage for thermal decay. The rate of heat production in high-level wastes after 5 years of storage is about 0.6% of that encountered one day after the termination of fission in the reactor. Sr-90 and Cs-137 are the two principal abundant isotopes decaying at this point in time. Wastes during this early period of storage have been and continue to be kept as aqueous solutions/slurries in tanks, and this method appears to be generally satisfactory for short-term storage. There are a few design features, however, which deserve considerably more attention:

a. The tanks should be so located and constructed that they will be secure from rupture by surface detonation of thermonuclear weapons;

b. They should be located in or on dry, highly impervious rock as remote as possible from active faults, mobile underground water, and aquifers;

c. The tanks should be designed for thorough decontamination in the case of damage or deterioration and should possess multiple containment barriers, each one of which can be decontaminated and repaired;

d. If left unattended, the tanks and secondary barriers should secure the contents for at least 500 years against the usual ravages of time.

It seems advisable in any case for the initial storage in water to be as short as possible. Water is a highly mobile fluid and requires a strong structure for containment; the chemical reactivity of aqueous solutions/slurries of intensely radioactive wastes is high. It should be possible to remove water after 1 to 3 years and reduce the waste to some suitable stable solid form. The form may have to be granular initially to permit better heat dissipation, but clearly it can be converted to a bulk glass or ceramic after 3 to 5 years. Studies by the AEC suggest that wastes may be solidified in granular form immediately after fuel reprocessing (i.e., after 0.3 years), thus obviating tank storage.[1,9] This research should be pursued vigorously.

8.7.2 Disposal in solar orbit. By the year 2000 it is expected that the entire U.S. nuclear industry will be producing on the order or several thousand metric tons of high level waste products per year. It is conceivable that this material could be placed in solar orbit by a group of some tens to 100 large boosters or by some alternate series of space station shuttles and interplanetary boosters.[6] This plan is both expensive - costing perhaps $5-10 billion per year by 2000 - and currently unfeasible. The probability of failures which could lead to atmospheric reentry must be essentially zero and, in addition, payloads must themselves be designed to survive reentry and impact in an intact and recoverable state. The vaporization of even 10^6 Ci of long-lived high-level waste products in the upper atmosphere would be a serious event leading probably to significant radiation doses to the lungs of man and animals. Our only experience with the reentry of radioactive material involved the disintegration in April 1964 of a SNAP thermoelectric satellite which contained 17,000 Ci of Pu-238 (about 900 g). It is estimated that this relatively small release of radioactivity will result in a cumulative 50-year dose to the lungs of adult humans of about 36 mrem (about 1/3 of one yearly dose from background radiation).[15]

The idea of disposal in solar orbit is satisfying in its simplicity and finality and should be reviewed periodically in light of developments in space technology.

8.7.3 Deposition in oceanic trenches. Oceanic trenches occur at lines along which one oceanic plate is sinking beneath the adjoining plate into the mantle of the earth. Ocean sediments accumulate to great depth in trenches and subside, in part at least, with the sinking plate, which may descend at rates of the order of 1-10 cm/yr. Thus, anything deposited in the sediments of the trench must become more deeply embedded by virtue of mechanical locking with the descending plate and the continuing deposition of sediments, and must eventually be ingested into the mantle of the earth.

An objection to the disposal of wastes in oceanic trenches arises from the fact that the radioactive wastes might be exposed to an aqueous environment - seawater and/or wet sediments for long periods of time during which they will still generate significant amounts of heat. The possibility of radiolytic decomposition of containers and the possible loss of material by accidental sinking away from the trench introduce considerable risk, even for the disposal of fission products. The method is appealing in that confinement forever is guaranteed, providing only that the method of initial implantation totally and irreversibly separates the radioactive material from the oceanic environment. This method should be reviewed periodically.

8.7.4 Disposal in Antarctica. Ice, like salt, exhibits plastic flow at modest pressures and is self-sealing. This property and the inaccessibility of the antarctic interior prompted Philberth[16] to propose disposing of high-level wastes deep in the Antarctic ice cap, probably at the ice-rock interface. The wastes, in an encased vitreous form, with dimensions 0.3 x 3 meters, and heat output of about 5400 watts, would be allowed to melt their way into the ice and slowly sink to the ice-rock interface.[17] It would take an estimated 3-5 years to sink an expected 1.5 km to the interface assuming an ambient ice temperature of about -50°C. On the high central plateau around the Pole of Inaccessibility, the temperature at the ice-rock interface has been estimated[18] (but not as yet directly measured) to be -20 to -30°C. This region therefore was assumed to have no free water at or near the interface, but this assumption is now in doubt, as will be discussed below. The absence of free water is crucial to the plan since there must arise no possibility of channels of communication between the deep interior ice and the sea, some 600-1000 km away.

The ice of Antarctica appears to be very stable and ancient. Data on ice cores from "young" ice taken from an active ice stream near Byrd Station suggest an ice age of about 100,000 years at depths of 2.2 km. Other studies show that the average temperatures in Antarctica have remained below freezing for more than 1 million years. On the other hand, some believe that very extensive lateral flows or "surges" of ice have occurred roughly every 100,000 years during the Pleistocene age. These would be expected to thin the ice cap considerably and forcefully scour the continent. Material from the ice-rock interface of the interior might be transported to the sea rapidly in a surge. Sites would be selected according to rock topography so that transport would be minimized in the event of a surge. In any case, the enormous heat of fusion of the ice sheet alone implies a general stability for the future. Antarctic ice contains a heat of fusion equivalent to the entire solar input to earth over about 100 years.

The repository area could probably cover some 40,000 km^2, so the waste containers could be deposited sparsely to avoid a general heating of the ice. One 5400-watt container per km^2 would contribute about 10% of the local geothermal gradient (about 1.2×10^{-6} cal/cm^2 per sec). It is possible that permafrost extends deeply into the rock in central Antarctica - perhaps to depths of 0.5-1 km - so that there is essentially no mobile ground water. Perhaps a safer alternate plan would be to emplace waste containers deep into the permafrost rock and effectively decouple them from all possible future motions of the ice.

Recent observations[19] raise serious objections to proposals to dispose of high-level radioactive wastes under the Antarctic ice cap. Radioecho soundings of the ice cap from aircraft have revealed some 17 regions, located at the ice-rock interface under ice 2.3-3.7 km deep, that exhibit strong specular reflections and smooth surfaces. These regions are interpreted as water pockets or "lakes" a few km in extent and minimally about 1 m deep. True depths or maximum depths are unknown. The existence of liquid water in bulk beneath thick ice in the central regions of Antarctica suggests that temperatures may be much closer to the melting point than heretofore believed. Additional heat loading at the interface, as would be caused by radioactive wastes, would now seem ill-advised. It is possible that interconnections exist among the lakes and even between the lakes and the sea.

The details of the ice-rock interface in Antarctica deserve to be studied further. If the radio-specular surfaces observed prove to be artifactual or if the existence of the bulk water is extremely atypical and related to strictly local heat sources, then Antarctica may be a suitable repository for fission products. Security for 800 years may be possible. I recommend in any case against placing wastes containing actinides in ice.

The difficulties of safely transporting wastes to Antarctica should be assessed carefully, but this lies beyond the scope of this book. The circumpolar region between about 65° and 80° south latitude experiences violent storms with great frequency. Even very modern ships have been disabled or sunk in heavy seas around Antarctica, and aircraft have been damaged and blown off course. Finally, the international character of Antarctica represents a problem, since disposal could be permitted only if broad agreement were reached among many nations, including the Soviet Union.

8.7.5 Deposition in underground caverns created by the detonation of nuclear devices. This proposal[20] involved the introduction of fresh high-level wastes as aqueous solution/ slurry into a cavern formed 1000-2000 m underground by the explosion of a nuclear device. Rock is a poor conductor of heat, and, initially, self-distillation - perhaps into surface heat exchangers - would serve to cool the wastes. Finally, after sufficient decay, the material would be allowed to dry and finally to fuse with the rock to distribute the wastes within a natural silicate matrix. The hazards of this method lie first in the extensive and unpredictable fracturing of rock in the vicinity of an underground nuclear explosion - even to the surface. There seems to be no way to assure the absence of channels of communication between the cavern and the surface or underground aquifers. The explosion may serve also to release geological strain locally. Thus, minor faulting may occur out to a distance of from one to a few km from the site and offer additional channels. The advantage is that the method is quick, inexpensive (perhaps 0.01 mills/kw-hr), and can accommodate fresh waste directly from the reprocessing plant. If each reprocessing plant could be supplied with such sites, the transportation of high-level wastes could be obviated. But, on balance, the plan seems very much more hazardous than others.

<u>8.7.6 Burial in ancient salt formations or dry rock at intermediate depths</u>. The method of disposal in bedded salt has been the subject of NAS studies,[21] and the AEC has carried out extensive tests in the late 1960s at a salt mine at Lyons, Kansas.[7,22,23] The 100 m thick formation is of Permian origin about 250 million years old and lies at a depth of 300 m. The plan involves the deposition of radioactive waste in the form of metal-encased ceramic cylinders, about 25 cm in diameter and 3-5 m long, into thick salt strata located at depths of the order of 300 m. The rooms would be back-filled with salt and sealed. Since water is the chief enemy, the salt deposits have to be ancient (have been secure

from water for eons), currently dry, bedded under impermeable layers of clay or shale, and they must lie in geologically quiet regions essentially free of active faults.

In addition to these geological requirements, the integrity of the deposits must not be violated or jeopardized by human activities. Such was not the case at the Lyons, Kansas, site and in 1972 the AEC was forced to abandon it as unsuitable for waste storage, despite the fact that it had just previously requested $25 million from Congress to proceed with actual waste burial. In its haste to proceed with disposal the Commission had both failed to perform an adequate geological survey of the site and had allowed the American Salt Company to continue with large-scale underground solution mining within a very short distance of the site.[24] The accidental discovery of old gas and oil bore holes penetrating the Lyons mine and the underground loss of a hundred thousand gallons of water from mining operations of the salt company, both of which might have led to a flooding of the repository, forced the AEC to abandon the Lyons site altogether.

Completely suitable sites may be difficult to find. A recent preliminary survey by the Kansas Geological Survey for the AEC[25] indicated that only one of several bedded salt regions in west and central Kansas seemed good enough to warrant further consideration. In fact, there may be very few suitable sites in the United States. While bedded salt deposits are extensive, those that are at least 70 m thick and located within 700 m of the surface occur at only few locations: most extensively in Kansas, two smaller areas in Michigan, and one small area in New York, south of Lake Ontario.[1] To these should be added, however, salt-potash deposits in southern New Mexico. Recent AEC studies suggest, tentatively at least, that these deposits are also suitable.

Salt has desirable characteristics: It will undergo plastic deformation from modest overburden pressures and is thus self-sealing even at depths of 300 m; it is a better heat conductor than rock. Rather detailed calculations suggested that heat from radioactive decay would not have caused stresses sufficient to damage the overlying strata at the Lyons site. If valid, these calculations suggest that the entire output of high level waste in the United States through 2000 could be accommodated in about 100 acres of bedded salt. The cost has been estimated to be about 0.05 mills/kw-hr for disposal of solidified waste 10 years old.

The method seems quite promising for fission products, but not for transuranic elements. The depths proposed - of the order of 300 m - seem rather shallow to insure security for millions of years. Stability over millions of years past can provide little measure of future stability in the case of rather shallow salt beds.

Another type of salt deposit, salt domes or diapirs, occur in the Gulf coastal region and also in the Arctic coastal plane where eroded gypsum domes occur commonly on the barren northern islands. Salt domes have a cylindrical or inverted teardrop shape some 5-13 km long and terminate with a somewhat bulbous cap about 2 km in diameter. They arise as a result of the isostatic disequilibrium experienced by low-density bedded evaporite deposits that lie at depths of some 10-13 km and were deposited, in the Gulf region, about 180 million years ago. Salt domes behave much like liquid bubbles, slowly pressing their way through overlying rock strata. Many domes lie at easily accessible depths of 1-3 km, and the occurrence of plastic flow assures a self-sealing, water-tight situation.

Salt domes would appear to be ideal structures for the disposal of solidified fission products, although the usual caveat must be offered regarding actinides. The chief problem with salt domes, however, is the fact that they are nearly always associated with valuable minerals. As the salt columns rise, sediments are pierced and steeply tilted along their flanks. Complex radial faulting occurs as well as block faulting directly above the dome. These effects tend to trap petroleum and gas against the flanks of and above salt domes. In addition, salt domes have a characteristic cap-rock structure which contains elemental sulfur of commercial importance.

Similar proposals have been made for storage and/or disposal in rock formations which are dry, impervious, show evidence of long geological stability, and are certain to have no future importance as mineral resources.[1,4,7,21] Ancient granites and metamorphic rock of the stable continental cores, some of which have radiological ages of 1 billion years or more, are possible candidates. Silicate rock has an advantage over salt in that it is essentially insoluble. On the other hand, rock does not undergo plastic deformation, at least at modest depth, and so is not self-sealing. The rigidity of rock, however permits mechanical excavations to depths of about 3000 m. Costs have been placed at about 0.07 mills/kw-hr for hard-rock options.[4]

As mentioned above, proposals to deposit liquid and slurried wastes in underground caverns have been considered by the AEC for at least the Hanford and Savannah River facilities.[9,13] In the latter case, the plan has been under study for some 10 years. A report by the NAS recommended discontinuation in 1966, but a NAS review panel concluded the plan had sufficient merit to justify construction of a shaft and tunnels. More recently, further negative opinion has been received from NAS. A major objection concerns the mobility of the water initially associated with the wastes and the possible mobility of ground water in the vicinity of the caverns.

Care must be exercised in the matter of thermal stresses on rock as the result of radiogenic heat from the wastes. In the vicinity around a source of heat in rock there will be a region of compressive stress merging into a region of relative tensile stress, and these regions will migrate as the thermal pulse slowly moves outward. Rock is very strong in compression but weak in tension. It seems advisable therefore that the heat loading be adjusted so that relative tensional stresses never exceed the static compression of the overburden, i.e., the net tension is never developed. If thermal fracturing of overlying rock can be avoided, the proposal to place wastes in stable rock formations seems attractive. Depending on rock characteristics, it may be possible to permit peak temperatures high enough to eventually fuse the wastes into the rock.

8.7.7 Deep burial. Though rock at normal pressure and temperature does not deform plastically, it will do so at the more severe conditions encountered at great depth.[6] The geothermal temperature gradient is about 25°C/km in the upper lithosphere. At 20 km, the temperature is of the order of 500°C at a pressure of about 5000 atmospheres. Under these conditions, slow plastic deformation into a cavity can occur. If it were possible to drill very deep holes into the stable silicate core of the continent, then ceramic cylinders of waste could be emplaced in a vertical array into a region where self-sealing and permanent entrainment would occur. The axial geometry also would be a favorable one for the dissipation of radiogenic heat. The plan is not now feasible and, in any case, probably would be expensive. If electrically heated, rock melting drills were used, I estimate a cost of perhaps $20 million per hole. Assuming a 2 km depth of cylinders, each hole could accommodate perhaps 500 cylinders, each with diameter about 25 cm and length of 4 m. Each 1 G w(e) plant will produce waste equivalent to about 8 cylinders per year (about 5 metric tons of waste products). Thus, one hole should prove adequate on average for the operational lifetime of each reactor. By the year 2000, about 16 holes/year will be needed to hold the waste then produced. The cost of boring alone would be about 0.05 mills/kw-hr.

One advantage with the method is that such long retention is assured that transuranic elements may no longer represent a serious problem.

<u>8.7.8 Injection of wastes in liquid form into wells</u>. It is a rather common industrial practice to dispose of various kinds of liquid wastes by pumping them via an encased well into porous rock formations such as sandstone or limestone. For example, U.S. Steel Corp. injects acid waste from steel-treating into a 1400 m well at Gary, Indiana, at a rate of some 6.5 million gallons per month.[5] In general, care is taken to select formations in which the porous rock stratum is layered between impermeable strata such as shale. The wastes can travel considerable distances laterally and examples are known in which capping strata that had been considered impermeable have either leaked or fractured as the result of hydrostatic effects.

According to an NAS-NRC report,[10] the following minimum conditions should apply if the injection of liquid wastes into porous formations is not to be possibly hazardous: the porous formation must not lie above, within, or below any formations which may contain potential mineral resources (oil, gas, salt, etc.) or fresh water. There should not be competing disposal or withdrawal operations within 100-300 km in the same formation. Vertical cracking of the overburden and failure of bore casings must not occur. There must be a thorough exploration of the formation, probably over many years, using tracers to determine the actual lateral rate of flow of injected liquid. The formation must be bedded continuously between impermeable clay or shale.

In the cases of limited reservoir capacity, low permeability, or viscous liquids, it is common practice to inject liquids under high pressure in order to fracture the rock, thereby increasing flow rate or capacity. This process is called "hydrofracturing." The behavior of such wells is often unpredictable, and surface contamination can occur, as was the case in a spectacular fashion in 1968 at the Hammerhill Paper Co. in Erie, Pennsylvania. There is also evidence, from the 4000 m well at the Army's Rocky Mountain Arsenal, Denver, that high pressure injections can trigger numerous minor earth tremors.

Shale fracturing is a waste disposal process currently in use at Oak Ridge National Laboratory for the permanent disposal of locally generated intermediate-level waste solutions. The practice began in 1966 and has so far disposed of about 450,000 Ci of Cs-137 and 30,000 Ci of Sr-90. In the process the waste solution is mixed with a blend of cement and other additives and the resulting grout is injected into impermeable shale formations at depths of 230-300 m (well below the level at which bulk ground water is encountered). The injection pressure propagates a thin horizontal crack in the shale which results in a grout plane 2-10 mm thick and 100-200 m across. The grout soon hardens into a solid form from which radioactive elements could be leached only very slowly. While shale fracturing may be acceptable for intermediate-level wastes, the geological uncertainties involved make its use in high-level waste disposal rather unlikely.

Overall, it would appear that methods involving the direct injection of liquids into geological formations do not represent acceptably safe ways to dispose of huge quantities of high-level wastes.

8.7.9 Long-term storage in surface structures. Essentially two arguments exist for long-term storage of solidified waste in surface vaults:

a. If only fission products are stored, then the periods involved (800 years) can be within the life expectancy of man-made structures. Witness the pyramids of Egypt.

b. Storage in surface vaults will buy time to develop satisfactory and totally safe methods of disposal.

The latter argument, while reasonable, will impose an expensive step between liquid tank storage and final disposal. The first argument above is acceptable only if structures surely can be built to survive not only the ravages of time but also the impact of nuclear weapons and social change. The formidable design characteristics must be similar to those outlined above for liquid waste storage-tanks. Foremost, they should secure their contents from accidental release even in the absence of human attendance. It is also desirable that they be impregnable to forcible entry. Even the pyramids have not fulfilled this latter criterion, and it is especially difficult to see how forcible entry could be prevented if the structures were, like the pyramids, left unattended.

If suitable structures can be built and if the transuranic elements can be eliminated from high-level wastes, then long-term storage in vaults offers the single advantage of retrievability. Surface vaults are clearly unsuitable in a world fraught with social and political instabilities over a scale of hundreds of years.

It is interesting that the AEC rejected years ago the idea of long-term storage of solidified high-level wastes in man-made vaults or mausolea. The focus for 15-20 years has been on disposal in geological formations. The idea has been resurrected recently since the salt mine at Lyons was abandoned around 1972 as a possible federal waste depository. In May 1972 the AEC announced new plans for the interim management of solidified wastes.[1] These plans call for the construction of "engineered surface facilities" (vaults) capable of storing the solidified wastes from commercial operations. These facilities are to be available for use in 1979-1980 when the first shipments of waste are ready for delivery from reprocessing plants. The facilities are to be designed to isolate the wastes safely for centuries with complete retrievability. The plan still stands at this writing, as far as can be determined. It strikes me as an expensive proposal born of desperation. It would appear that a permanent disposal plan cannot be assured by 1980.

8.7.10 Massive environmental dilution. While not seriously considered by the AEC, the idea of dilution as a means of disposal nevertheless appears now and then, often in subtle forms. For example, Weinberg[6] wrote regarding salt mines:

"The mine would contain 38 tons of Pu-239 mixed with about a million tons of nonradioactive material. The plutonium in the cans is thus diluted to 38 parts per million; since plutonium is, per gram, 10,000 times more hazardous than natural uranium in equilibrium with its daughters, these diluted waste materials would present a hazard of the same order as an equal amount of pitchblende. Actually, the 38 tons of Pu-239 is spread over 200 acres. If all the salt associated with the Pu-239 were dissolved in water, as conceivably could result from total flooding of the mine, the concentration of plutonium in the resulting salt solution would be well below maximum permissible concentrations. In other words, by virtue of having spread the plutonium over an area of 200 acres, we have to a degree ameliorated the residual risk in the most unlikely event that the mines are flooded."

If the described situation is innocuous, then a simple procedure would be to dilute wastes directly into the oceans of the world. The oceans contain quite enough water to dilute the present and anticipated wastes to levels permitted for drinking. For example, after 100 years of land storage, only 0.08 km^3 of water is required to dilute the waste actinides in one metric ton of used reactor fuel to the Radiation Concentration Guide (RCG) values.[26] Only 23 km^3 of water would be required for the total wastes in the same one ton of fuel. The reason this is not done is quite obviously the fact that the biosphere tends to concentrate certain of the fission products and actinides. Concentration factors of 10^5-10^6 are not unusual and may be realized over one to a few trophic levels. The important consideration in all such discussions is the appearance of radioactivity in man and the biosphere rather than in the inanimate world per se. We feel that emphasis upon RCG values is often misleading and should not substitute for estimates of sequestration in the biosphere. This is illustrated abundantly in Chapter 7 of this book.

A more interesting approach to disposal through dilution was alluded to above, namely, the possibility that wastes may be so diluted <u>within some inert solid matrix</u> that their hazard becomes commensurate with that encountered naturally through the continental abundance of uranium, potassium, thorium, and radium.[27] The mineralogical background represents an irreducible minimum hazard. Suppose, for example, that the strip mining of coal were combined with the disposal of radioactive wastes. Each unit of waste would be intimately mixed with vast amounts of finely divided overburden to dilute the waste to a level something akin to the radiological hazard of granite. Finally, the mixture would be reconsolidated (probably melted) into a rock-like form to prevent leaching of the radioactivity by acidic or salt water. This process clearly would be both expensive and cumbersome, and it can be dismissed as a practical method of disposal.

<u>8.7.11 Transmutation and fission of transuranic actinides</u>. The various proposals discussed above require storing wastes on or in the earth for whatever times are needed for decay to nonhazardous levels. Inherent in these plans is the idea that men must bear the responsibility for actions over times that are long compared with the lives of nations. The requisite social and legal institutions simply do not exist to assure responsibility over periods of 800 years (fission products) upward to millions of years (actinides). <u>It</u>

seems prudent, therefore, to do everything possible to minimize the time required for the wastes to decay to safe levels. The one process which will decrease containment time by more than 3 orders of magnitude - from several million years down to 800 years - is nearly total removal of transuranic elements.[4,26]

At present the recovery of uranium and plutonium from fuel reprocessing is 99.5% and most of the other actinides are not recovered. This compares with a technological limit for extraction of actinides from the waste stream of about 99.9999% (actinium through plutonium) and 99% (americium through einsteinium). As can be seen in Figure 8.3, such complete extraction would lower the long-term relative toxicity by 3 orders of magnitude compared with current practice. Extraction of 99.9-99.95% uranium, neptunium, and plutonium and of 99% americium and californium (middle curve) can be accomplished realistically and would result in a long term actinide hazard about equal to that for fission products.

If the recovered actinides were burned up by recycling through a light water reactor, the additional costs for improved extraction plus burn-up would be about 1.5 times that for current fuel reprocessing and result in an increase[4] in delivered electricity of about 0.3 mills/kw-hr on a base of 20-25 mills/kw-hr. It is believed that light water reactors would not have to be altered much, except to increase U-235 enrichment slightly, in order to burn up actinides by fission. A much better route would be to burn up actinides in fast neutron reactors where the fast neutron spectrum is more efficient than thermal neutrons in transmuting actinides to fissionable iso-

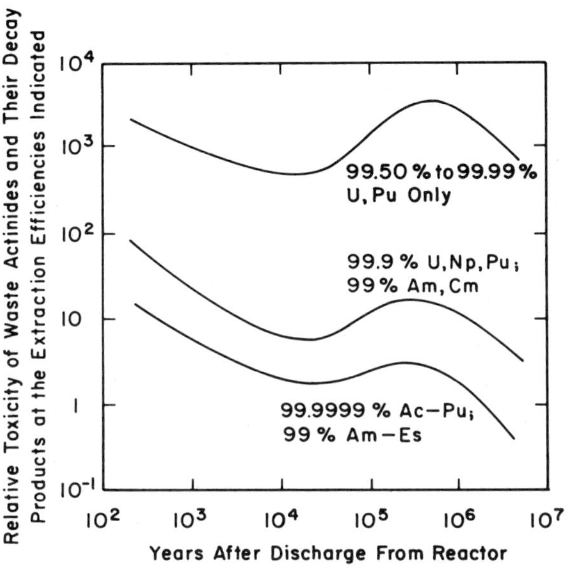

Figure 8.3 Relative toxicity of waste actinides and their decay products for different degrees of actinide extraction. Burn-up of fuel is as in Figure 8.1. Top curve represents current practice (99.5% uranium and plutonium); middle curve represents practical recoveries; lower curve represents likely limits. Adapted from ref. cited in note 4.

topes. If fast neutron reactors were used, the cost increment on electricity is estimated to be only 1% or less for improved extraction and burn up. At steady state, about 1 part in 7-10 of the fuel in a plutonium-fueled liquid metal fast breeder reactor could be actinides other than plutonium.

Geological burial or other disposal schemes would still have to apply to the fission products. The increment of cost cited above, therefore, is an excess of the cost of disposal of fission wastes, the costs of which may be expected to lie in the range 0.005-

0.008 mills/kw-hr.

The alternative is the continued production of waste material so laden with plutonium and other long-lived transuranic elements that confinement periods of millions of years will be necessary. While secure storage or burial may well be possible for periods of 800 years, this is not necessarily the case for periods thousands of times this. Within one million years, another ice age could occur, continental erosion of hundreds of meters is likely, and even the species man could become extinct.

If the transuranic elements were removed almost completely, then _storage_ in massive surface vaults for 800 years or any one of a number of deep burial schemes - e.g., in ancient salt deposits - becomes feasible. If the problem is one of the _disposal_ of transuranium elements, then surface vaults are excluded and some schemes for geological burial, including the now-favored burial in salt mines, become of questionable security.

Appendix 8.1. The Leak of 8 June 1973 from Carbon Steel Tank No. 106T, Hanford Facility, Richmond, Washington

The loss of some 115,000 gallons of liquid waste from tank 106T was discovered on 8 June 1973 and reported shortly thereafter (Seattle Post Intelligencer, Wed., 13 June 1973, p. 1; Atomic Energy Clearing House 19, #25, 18 June 1973). The tank is of 533,000 gallon capacity and contained prior to the leak about 400,000 gallons. The leak represents the second largest release of radioactivity of any leak - some 40,000 Ci of Cs-137 and some 14,000 Ci of Sr-90 - but was the largest of any leak by volume. Previous leaks totaled 307,000 gallons and involved some 200,000 Ci of Cs-137 and lesser amounts of other isotopes.

Tank 106T was built in 1944 and so represents one of the older tanks among the approximately 150 at Hanford.

The sheer volume of the leak raises important questions regarding the detailed distribution of radioactivity now in the ground beneath the tank. The bottom of the tank lies some 18 m below ground level, and the tank seems to have failed at or near its bottom where corrosion damage would be expected to be most severe. The injection of 20% of a tank volume into earth with an average porosity of 15-20% would be expected to bring radioactivity to a depth of 33 m just through filling the void spaces. The depth of penetration could be greater due (a) to the fact that the downward migration of liquid may continue even after lateral migration has slowed or stopped; (b) the ground is likely to be inhomogenous and to contain fissures or highly porous channels which together would provide avenues for deep penetration; (c) a fraction of the void spaces will resist filling in the case of a rapid leak and may be by-passed. The water table beneath tank 106T lies at a depth of about 70 m. On the basis of the above it is likely that significant amounts of radioactivity (more than 1000 curies) may have come close to the water table.

This possibility seems to be substantial from drilling data as summarized in "Report on the Investigation of the 106-T Tank Leak," July 1973, Hanford Facility. While radioactivity from the leak has not yet been detected in ground water beneath the tank, the downward penetration was at least to about 25 m below the tank or about 43 m from the surface. Whether radioactivity penetrated further at isolated points beneath the tank is not entirely clear. The maximum lateral migration is believed to be about 28 m from 106T.

The newspaper article of 13 June suggested that the leak was detected by sampling in a monitor well. Telephone interviews with officials at Hanford indicated, however, that the leak may have been detected by a drop in liquid level during a routine check. Both methods are crude and are useful only in detecting massive leaks. Neither gives an immediate notice of the start of a leak and neither seems useful in detecting persistent small leaks. The leak from tank 106T, for example, is now known to have started about 7 weeks prior to June 8, 1973.

The possibility exists, therefore, that other tanks at Hanford and elsewhere may be leaking and may remain undetected. (The UCS has yet to discover what exactly are the AEC criteria which determine whether an old tank is fit for service.)

If old tanks can fail massively under static conditions, it seems likely that even a minor earthquake could initiate a number of massive and near simultaneous leaks. While Hanford is located in what seems currently to be seismically, a rather quiet region, there are faults in the vicinity which have undergone displacement in geologically recent times.

Appendix 8.2. Text quoted from page 1 of ref. 11: USAEC-WASH-1520, Contaminated Soil Removal Facility. Richland, Washington, April 1972

"The U.S. Atomic Energy Commission plans to remove plutonium contaminated soil from the floor of an existing enclosed trench (Z-9) used between July 1955 and June 1962 as a subsurface disposal facility for plutonium contaminated liquids from the Plutonium Finishing Plant on the Hanford Reservation near Richland, Washington. It is estimated that the soil to be removed contains approximately 100 kilograms of plutonium in a volume of approximately 1800 cubic feet.

"Liquid wastes from the Plutonium Finishing Plant (PFP) have been discharged to subsurface disposal facilities (enclosed trenches) since startup of the facility approximately 22 years ago. Most of the plutonium contained in these liquid wastes is sorbed (retained) by the soil and held within a few feet vertically of the point of release. The enclosed trenches are located in a fenced off area well within the boundaries of the controlled Hanford Plant. Careful surveillance using test wells has allowed this practice to be followed safely for 22 years. Due to the quantity of plutonium contained in the soil of the Z-9 enclosed trench, special precautions and emergency plans are required for Z-9 which are not required for other enclosed trenches.

"It is proposed to construct facilities at Z-9 to permit excavation and packaging of contaminated soil, to add equipment to the existing Plutonium Finishing Plant to permit recovery of plutonium from the contaminated soil, and to construct an Underground Storage Vault fourteen feet wide by eight feet high by 400 feet long for interim storage of contaminated soil.

"Removal of the plutonium contaminated soil will eliminate the need for special precautions and emergency plans necessary to assure the safe storage of the plutonium in the enclosed trench. Due to the quantity of plutonium contained in the soil of Z-9 it is possible to conceive of conditions which could result in a nuclear chain reaction. These conditions would be the rearrangement of the contaminated soil, flooding of the enclosed trench following a record snowfall and rapid melting (Chinook), and failure to implement planned emergency actions (pumping of flood waters from adjacent terrain and addition of neutron absorbing materials to the enclosed trench). Even though the probability of all these occurrences happening in sequence is extremely remote, removal of the Pu contaminated soil will eliminate any possibility of such an event."

References

1 WASH-1250, The Safety of Nuclear Power Reactors (Light-Water Cooled) and Related Facilities, USAEC, Washington, D.C., Draft, December 1972.

2 S. Novick, Environment 13, No. 9, 42 (1971).

3 C.H. Fox, Radioactive Wastes, USAEC, Div. of Tech. Information, Washington, D.C. 1969.

4 A.S. Kubo, Technology Assessment of High Level Nuclear Waste Management, ScD thesis, Dept. of Nuclear Eng., Mass. Institute of Technology, Cambridge, Mass., 1973. This work is summarized in part in a paper by A.S. Kubo and D.J. Rose, Science 182 1205 (1973).

5 H.P. Metzger, The Atomic Establishment, Simon and Schuster, New York, 1972.

6 A.M. Weinberg, Science 177 27 (1972).

7 ORNL 4451. Siting of Fuel Reprocessing Plants and Waste Management Facilities, Oak Ridge National Laboratory, published by National Tech. Info. Service, July 1970.

8 SCEP Report, Man's Impact on the Global Environment, MIT Press, Cambridge, Mass., 1970.

9 WASH-1202, Plan for the Management of AEC-Generated Radioactive Wastes, USAEC, Washington, D.C., January 1972.

10 Untitled report to the Division of Reactor Development and Technology, USAEC, from the

Committee on Geological Aspects of Radioactive Waste Disposal, Division of Earth Sciences, NAS-NRC, May 1966.

11 WASH-1520, Contaminated Soil Removal Facility - Richland, Washington, USAEC, Washington, D.C., April 1972, Environmental Statement.

12 Letter of 2 March 1973 to D.L. Ray, Chairwoman, USAEC, from John E. Bryson and Terry R. Lash of Natural Resources Defense Council, Inc., Palo Alto, Calif.

13 J.W. Bingham, C.J. Londquist, and E.H. Baltz, Geological Investigation of Faulting in the Hanford Region, Washington, U.S. Department of the Interior, Geological Survey, Tacoma, Washington, 1970.

14 WASH-1528, Future High Level Waste Facilities - Savannah River Plant, Aiken, South Carolina, USAEC, Washington, D.C., December 1972, Draft Environmental Statement.

15 B. Shleien, Health Phys. 18 267 (1970).

16 B. Philbert, Schweizerische Zeitsch. f. Hydrologie, Fasc. 1 (1961), p. 262.

17 E.J. Zeller, D.R. Saunders, and E.E. Angino, Bull. Atomic Scientists 29, No. 1, 4 (1973).

18 W.F. Budd, D. Jenssen, and U. Radok, Meteorology Dept. Publ. No. 18, University of Melbourne, 1971.

19 G.K.A. Oswald and C. deQ. Robin, Nature 245 251 (1973).

20 Cohen et al., Nuclear Technology April 13, 1972, p. 76.

21 Disposal of Solid Radioactive Wastes in Bedded Salt Deposits, NAS-NRC, Washington, D.C. 1970; Disposal of Radioactive Wastes on Land, Publication No. 519, NAS-NRC, Washington, D.C., 1957.

22 ORNL-4555, R.L. Bradshaw and W.C. McClaim, eds., Project Salt Vault: A Demonstration of the Disposal of High-Activity Solidified Wastes in Underground Salt Mines, Oak Ridge National Lab., Published by Natl. Tech. Info. Service, April 1971; ORNL-4751, A. Bock, Radioactive Waste Repository Project: Technical Status Report for Period Ending Sept. 30, 1971, Oak Ridge National Lab, published by Natl. Tech. Info. Services, December 1971.

23 R.D. Cheverton and W.D. Turner, Power Engineering April 1972, p. 48.

24 W.W. Hambleton, Technology Rev. March/April 1972, p. 15.

25 The report from the Kansas Geological Survey to the USAEC was announced in AEC press release of January 21, 1972. The Discussions and Conclusions section (Chap. VII) of the report, along with a map showing the salt beds surveyed, were reprinted in the release.

26 ORNL-TM-3964. H.C. Claiborne, Neutron Induced Transmutation of Transuranium Elements,

Oak Ridge National Lab., published by Natl. Tech. Info. Service, December 1972.

27 J.H. Rubin, High-Level Radioactive Waste Management in the U.S., IAEA Panel to Examine the Possibility of Establishing International Storage Sites for High-Level and Alpha Bearing Radioactive Wastes, November 20-24, 1972, Vienna.

28 O.R. Frisch, ed., The Nuclear Handbook, Van Nostrand, New York, 1958.

29 F.K. Pittman, "Management of Commercial High-Level Radioactive Waste," a paper presented at a summer course on Nuclear Fuel and Power Management, MIT, Cambridge, Mass., 25 July 1972.

Index